ENERGY SUPPLY

LIBRARY IN A BOOK

ENERGY SUPPLY

Lisa Yount

Facts On File, Inc.

ENERGY SUPPLY

Facts On File, Inc.
132 West 31st Street
New York NY 10001

Library of Congress Cataloging-in-Publication Data

Yount, Lisa.
Energy supply / Lisa Yount.
p. cm. — (Library in a book)
Includes bibliographical references and index.
ISBN 0-8160-5577-7
1. Power resources. 2. Energy consumption. I. Title. II. Series.
TJ163.2.Y68 2005
333. 79—dc22 2004021607

Facts On File books are available at special discounts when purchased in bulk quantities for businesses, associations, institutions or sales promotions. Please call our Special Sales Department in New York at (212) 967-8800 or (800) 322-8755.

You can find Facts On File on the World Wide Web at http://www.factsonfile.com

Text design by Ron Monteleone
Graphs by Jeremy Eagle

Printed in the United States of America

MP Hermitage 10 9 8 7 6 5 4 3 2 1

This book is printed on acid-free paper.

To my father, Stanley G. Yount,

who had more energy than anybody

CONTENTS

PART III
Appendices

PART I

OVERVIEW OF THE TOPIC

CHAPTER 1

INTRODUCTION TO ISSUES IN ENERGY

Energy is the power of change. It is the ability to do work, which means altering the position, structure, or chemical nature of materials or objects. Without energy, the Sun and stars could not shine, and life could not exist.

All energy is either actual (doing work at the moment) or potential (stored, able to do work when released). Forms of energy include nuclear, chemical, electrical, mechanical, and thermal (heat). Most energy on Earth ultimately derives from atomic reactions within the Sun, sent through space as electromagnetic waves and high-energy particles.

Capturing, storing, transporting, and using energy often involve changing it from one form to another. Engines can transform heat or mechanical energy into electricity, for instance. Energy can also be converted from potential to actual and back again. The transformation of energy, or thermodynamics, is governed by two basic laws. The first states that energy cannot be created or destroyed. The second states that the use of energy to do work is never completely efficient: Some energy is always lost as heat. This means that once energy has done work, it cannot be used to do the same amount of work again without adding more energy.

All living things capture, store, and use energy. Through the process of photosynthesis, green plants use solar (sun) energy to turn carbon dioxide and water into compounds that store and use chemical energy in their bodies. Animals consume these chemicals when they eat plants (or eat other animals that eat plants) and use the energy from them to carry out their daily activities. When plants and animals die, the energy stored in their bodies returns to their environment.

Early in their existence, humans began harnessing energy outside their bodies. They warmed and lighted their homes and cooked their food by burning wood, which released the energy stored in plants. They used the energy of animals for transportation. They used the energy of wind and water to turn mill wheels and sail ships. Their use of energy shaped their civilizations, and in turn, changes in civilization affected energy use. The more complex civilizations became, the more energy they used.

The higher a country's standard of living, the more energy each of its citizens uses. Not surprisingly, therefore, energy use is higher in the developed than in the developing world. The United States is one of the world's most energy-hungry countries, with a yearly consumption of 339 million Btus (British thermal units) per person—about five times the world average, and a quarter of the world's total consumption. The average citizen of Western Europe or Japan uses about half as much, and the average citizen of Africa less than a 20th.

In a book called *Small Is Beautiful*, published in 1975, E. F. Schumacher wrote, "There is no substitute for energy. . . . It is not 'just another commodity' but the precondition of all commodities, a basic factor equally with air, water, and earth."[1] Energy's importance is, if anything, even greater today. The drive to control natural, or primary, energy sources, especially oil and natural gas, increasingly shapes world politics and triggers wars. In secondary forms, such as electricity and the refined fuels that power internal combustion engines, energy underlies all economic activity. The choices people make about energy may even determine whether life on this planet survives.

SOURCES OF ENERGY

The ideal energy source would be limitless in supply, easy and cheap to extract and convert into usable form, efficient at doing work, and free of pollution or other environmental damage during both extraction and use. Unfortunately, no such source exists. All energy sources have economic and environmental advantages and disadvantages. Comparing these pluses and minuses is one of the hardest parts of evaluating the world's energy supply.

The energy sources used most today are the so-called fossil fuels: coal, oil (petroleum), and natural gas. In 2001, they were the source of about 86 percent of the energy produced in the world. They are called fossil fuels because most scientists believe they were made from living things that died between about 440 million and 5 million years ago. Instead of being decomposed by bacteria with the help of oxygen in the air, these remains sank beneath swamps, lakes, or oceans and were therefore cut off from oxygen. Layers of sand and mud slowly fell on top of them, squashing them into a thick soup of carbon and hydrogen molecules that seeped into pores in the rock forming around them. If the rock was pushed far enough down into the Earth's crust, high pressures and temperatures transformed the hydrocarbon soup into petroleum, often with a layer of natural gas above it.

In the late 20th century, concern about the world's heavy dependence on fossil fuels began to grow for several reasons. First, most experts say that these substances are no longer being formed in significant amounts and that most large deposits of them have been discovered. Although analysts disagree strongly about the exact date, many believe that worldwide production of fossil fuels will start to decline permanently during the first half of the 21st century.

A second problem is that the greatest concentrations of these fuels, especially of oil, are in politically unstable areas such as the Middle East. When Arab countries stopped shipping oil to the United States between October 1973 and March 1974, in retaliation for United States support of Israel in the Arab-Israeli "Yom Kippur War" of October 1973, Americans learned suddenly and painfully that they could not take this source of fossil fuels for granted. As fossil fuel supplies begin to dwindle, struggles for control of them are likely to contribute increasingly to international tensions and even wars.

Perhaps the most important reason for worry about fossil fuels, though, is the health and environmental problems that their extraction, transport, and use create. Burning them releases many pollutants, including sulfur dioxide, nitrogen oxides, and ash particles (particulates). Scientists at Harvard University have claimed that burning fossil fuels causes about 60,000 deaths in the United States each year, and the American Lung Association says that particles from smokestacks are contributing to the rapid rise in childhood asthma. Airborne sulfur dioxide and nitrogen oxides also combine with water in the atmosphere to produce so-called acid rain, which environmentalists say has damaged forests and killed fish in lakes.

The worst pollutant of all may be carbon dioxide (CO_2). All burning fossil fuels give off this colorless, odorless gas, along with water. Carbon dioxide is harmless in itself; humans and animals emit it with every breath, and plants need it to live. However, it is transparent to the short wavelengths of light that the Sun beams down onto the Earth but opaque to the longer wavelengths that radiate back into space. This means that the higher the concentration of CO_2 in the atmosphere, the more solar energy is kept close to the Earth's surface, and the warmer the surface becomes.

Climate scientists think that increases in atmospheric carbon dioxide and certain other gases, such as methane (the main component of natural gas), create a "greenhouse effect" that could raise the overall temperature of Earth by several degrees. Such a change might seem minor, but many researchers believe it would radically alter weather and climate patterns, causing droughts in some places and floods or violent storms in others.

Earth's mean surface temperature appears to have increased by 1.1°F during the 20th century, and examination of many sources of information about climate change, including tree rings, ice cores, and historical records, suggests that no comparable rise has occurred in at least a thousand years. At the same time, measurements have shown an increase of carbon dioxide in the atmosphere of about 0.5 percent per year since 1958. Most, though not all, experts think this rise is caused primarily by human burning of fossil fuels.

Scientists and politicians argue intensely about exactly how much warming has taken place, how much of it is due to human activities as opposed to natural climate variation, how much warming will occur in the future, what effects it will have, and what should or can be done about it. Some say that the Earth's climate system is too complex for current computer models or other technology to predict accurately. Most agree, however, that some warming is occurring and that increases in the atmospheric concentration of greenhouse gases caused by human activities are helping to cause it.

Because of these concerns about fossil fuels, growing numbers of people say that the world's energy should come as much as possible from other sources. Some sources, chiefly nuclear power and hydroelectric power (energy from moving water), have been in use for some time and, like fossil fuels, are classified as "conventional." Others, including sun, wind, hydrogen, burning organic matter (biomass), and heat energy from the Earth (geothermal power), are more experimental and are often called alternative energy sources. They are also sometimes referred to as renewable or "green" (minimally harmful to the environment). These terms can be confusing because they overlap but are not identical. Hydropower, for instance, is usually considered to be renewable but not alternative, and geothermal energy, as presently used, is alternative but not renewable. Alternative energy sources generally are easier on the environment than conventional sources, but none is completely harmless.

Despite their environmental advantages, renewable energy sources represented only about 7 percent of the world's primary energy supply in 2001—and almost all of that was conventional hydropower. The chief barrier to the use of other renewable sources is cost. Although the fuels themselves are essentially free, using them often requires technology that is still more or less experimental, and power plants or other facilities that employ them are more expensive to build than plants that use fossil fuels. Furthermore, many alternative energy sources are both diffuse (they require much more land per unit of energy than burning fossil fuels) and variable, and they do not occur in places and times where energy is most in demand. They therefore require expensive shipment, backup energy supplies, or both.

Supporters of fossil fuels claim that alternative fuels are not now and probably never will be economically competitive with fossil fuels in a free market. Alternative energy boosters, however, say that part of the cost disparity is due to the fact that the price of fossil fuels, especially in the United States, is kept artificially low. A 10-year study by the European Commission and the U.S. Department of Energy concluded in 2001 that the cost of electricity made from coal and oil would double, and natural-gas costs would rise by 30 percent, if so-called externalities, such as environmental and health damage caused by pollution from fossil fuel burning, were factored in. However, a report by the U.S. General Accounting Office stated that "electricity from renewable energy usually costs so much more than electricity from fossil fuels that externality considerations do not overcome the difference."[2] Critics of the externality idea point out that fossil fuel prices do not include the cost of meeting environmental and health regulations.

Each source of energy, conventional or alternative, has its own technology, history, uses, and advantages and disadvantages. The following sections provide a brief overview of these sources, beginning with the most common—fossil fuels.

COAL

Coal is solid carbon, combined with varying amounts of sulfur and other impurities. Its precursor is peat, compacted masses of plant matter that form beneath marshes and lakes. Peat turns into coal when it is further compressed and hardened into rock.

Coal is classified into several grades, depending on the amount of heat it gives off when burned and the quantity of impurities it contains. Anthracite, the highest grade, is more than 90 percent carbon. It is hard, shiny black, and produces heat with great efficiency. Bituminous coal and lignite, or "brown" coal, are lower grades. They exist in larger quantities and are easier to extract than anthracite but are less efficient and more polluting.

As with all energy sources, coal is distributed unevenly in the world. The largest proven deposits are in the United States, Russia, and China, which together contain about 60 percent of the world's known supply. The greatest coal reserves in the United States are in the East, especially West Virginia and Pennsylvania, and in the Rocky Mountains, especially the Powder River Basin of Montana and Wyoming. Eastern coal burns more efficiently than western (Rocky Mountain) coal, but it has a larger amount of sulfur and other pollution-producing contaminants, so western coal has been more in demand since concern about pollution increased.

Because coal is a sedimentary rock, it usually exists in layers or seams. Traditionally, workers dug shafts into seams and mined the coal from them by hand. Underground mining often did not disturb the landscape much, but it seriously threatened the safety and health of the miners. Conditions have improved, but underground coal mining is still dangerous and unpleasant work. Underground mines also often leave heaps of refuse containing poisons that can leach into water supplies.

Today, about two-thirds of the coal mined in the United States is collected by surface mining, which its critics call strip mining. Huge machines tear soil and rock from the earth's surface and then scoop out the coal. Sometimes whole mountaintops are razed, producing tons of debris that are dumped into neighboring valleys. Environmentalists and local residents have been extremely critical of the habitat destruction and visual blight this kind of mining causes. In response to these criticisms, Congress passed the Federal Surface Mining Control and Reclamation Act in 1977, which requires mining companies to restore the topsoil and vegetation to areas that they have finished using. The effectiveness of this law and of the restoration efforts it has produced is much debated.

Of the three fossil fuels, coal was the first to be used extensively. Europeans, especially Britons, began burning it to heat their homes in the 17th century. The steam engine that James Watt invented in 1765 burned coal in a boiler to create the steam, so the Industrial Revolution that grew out of his invention was built on coal. So was the settlement of the western United States in the second half of the 19th century, spurred in part by railroads that both burned and hauled coal. Coal was also heavily used in steelmaking and other industries during the late 19th and early 20th centuries. In developed countries today, however, most coal is burned in power plants to generate electricity. Coal provides 23 percent of the energy consumed in the United States, and more than 90 percent of that is used in electric power plants. Coal generates a little more than half the country's electricity.

Coal has the advantage of being plentiful and cheap, at least in the United States. It is bulky and therefore hard to ship, however, and converting it into more convenient forms such as liquid or gas is expensive. Most important, it is the most polluting of fossil fuels. About two-thirds of acid rain is estimated to come from sulfur dioxide and nitrogen oxides spewed out by coal-burning power and steel plants. Burning coal also produces mercury, which can damage the nervous system. The federal Clean Air Act requires coal-burning plants to remove most sulfur from their coal either before or after it is burned, however, and coal supporters say that up to 90 percent of sulfur and ash can now be removed. A remaining problem is that, according to many critics, burning coal releases more carbon dioxide (CO_2) into the atmosphere per unit of energy than any other fossil fuel. The U.S. Department of Energy's Energy Information Administration

(EIA) says that coal accounted for 36 percent of the country's carbon dioxide emissions in 2001.

OIL

Since ancient times, people in some parts of the world have noticed a black, smelly goo seeping out of the ground. This material was raw petroleum, or crude oil. (*Petroleum* comes from Latin words meaning "rock oil.") It is made up of many kinds of hydrocarbons, or compounds of hydrogen and carbon. Unlike coal, petroleum exists naturally in liquid form, filling the pores of certain types of rocks as water fills the pores of a sponge.

The industry magazine *Oil and Gas Journal* estimated that the world had unextracted reserves of 1,213 billion barrels of oil in January 2003. The Middle East possesses more than half of it, with more than a fifth of the world's total lying in the country of Saudi Arabia alone. Non–Middle Eastern countries with large oil reserves include Canada, Venezuela, Nigeria, and Russia. Oil reserves in the United States—only about 5 percent of the world's supply—are concentrated in Texas, Alaska, Louisiana, and California.

Seeking underground oil was at first mostly guesswork, usually based on the presence of surface oil nearby, but 20th-century geologists have learned much about the types of rocks and formations in which oil is found. The most common formation, called an anticline, is a layer of porous, oil-containing rock trapped beneath a layer of dense, hard rock that is impenetrable to oil. The oil-bearing rock frequently sits on top of a lake of saltwater, with which oil also will not mix, and has a bubble of natural gas above it. Often the anticline is shaped like an inverted cup, which may show at the surface as a dome or hill. The dome keeps the oil from escaping and prevents oxygen from reaching and decomposing it.

Geologists have learned that a porous layer will contain oil only if it is, or at some time in its history has been, between 7,500 and 15,000 feet (2,272 and 4,500 meters) below the surface. If the layer has remained higher than this "oil window," organic material will not have been heated enough to break it down into petroleum. If it has been pushed lower, it will have encountered heat and pressure great enough to break down hydrocarbons into more or less pure methane (natural gas).

One of the most common methods used to spot potential oil deposits today is called seismic exploration. Much as bats and dolphins navigate by sending out high-pitched sound waves and analyzing the echoes produced when the waves bounce off objects in the environment, geologists send high-intensity soundwaves into earth or water and use supercomputers to interpret the echoes that come back. Between the 1920s, when seismic exploration was first used, and the 1950s, the sound waves came from dynamite explosions. Modern seismic surveying, however, uses hydraulic systems mounted on heavy trucks to generate the waves. For undersea exploration, ships shoot compressed air from air guns to generate the sound waves (which environmentalists say harms whales and other marine life) and pick up the reflections with long strings of towed hydrophones. The best seismic techniques produce three-dimensional diagrams of underground formations.

The Chinese began drilling for oil as early as 600 B.C., using bamboo poles and weights to make holes as deep as 1,500 feet. Men jumped up and down on the poles to drive them into the ground. Mechanical versions of this "spring pole" technique made 19th-century wells. The rotary rig, consisting of a rotating steel pipe with a toothed bit at the bottom, was introduced at the start of the 20th century. Recently, engineers have learned how to drill wells on a slant or even horizontally. This lets them drill multiple wells from the same rig, which both saves money and reduces impact on the surface environment.

Oil is stored under pressure, often with a layer of natural gas above it and a layer of salt water below. When a drill breaches a layer of oil-bearing rock, the pressure is released and the gas and water expand, pushing the oil to the surface—sometimes explosively, creating the gushers that caused so much excitement in the industry's early days. (Gushers, or blowouts, waste tremendous amounts of oil and also cause great environmental damage, so modern oil drillers try to prevent them.) Once enough gas is released to lower the pressure below the minimum needed to pull oil from the rocks, drillers add pumps to draw up the remaining oil. The bobbing, insectlike structures seen in many oil fields are one common kind of pump. Gas, water, or steam under pressure may also be injected into the well to drive the oil upward.

Most crude oil is shipped to refineries to be broken up into smaller molecules. The refineries heat the oil in a tall fractionating tower and collect the hydrocarbons distilled out of the resulting gas at various temperatures. Gasoline and other simple, lightweight molecules condense near the top of the tower; longer, heavier molecules called polymers are collected near the bottom. Because heating alone does not produce enough of the most desirable compounds, the refineries break up some of the larger molecules by a process called cracking.

In places where petroleum seeped out of the ground by itself, ancient cultures used it to waterproof boats, treat skin diseases, and even embalm mummies. They burned it in torches and, after distilling it, in lamps. The medieval Islamic civilizations developed a process for refining crude oil to make a liquid later called kerosene, which was a better lamp fuel than distilled oil. After a U.S. inventor named Michael Dietz created a cleaner-burning kerosene lamp in 1857, kerosene began to replace whale oil as a popular lamp fuel in Europe and North America as well. (Environmentalists often portray the oil industry as a villain, but its supporters say that—albeit inadvertently—it probably helped to save whales from extinction.) Nineteenth-century kerosene was usually extracted from coal at first, but people soon realized that oil was at least as good a source. Entrepreneurs saw that the gooey black stuff could be valuable.

A former railroad conductor named "Colonel" Edwin Drake ("Colonel" being a courtesy title) drilled the first U.S. oil well near Titusville, Pennsylvania. He struck oil at 69 feet below the surface on August 27, 1859. His success triggered an "oil rush" much like the gold rush that had swept California 10 years earlier, spawning a hundred more wells and multiplying the area's population many times over within a year. A similar rush followed the discovery of Spindletop, the first large oil field in Texas, in 1901. Opening with an explosive gusher that shot oil more than 150 feet into the air, Spindletop, near the town of Beaumont, at first produced almost 100,000 barrels of oil a day, more than the combined output of all the other wells in the country at the time.

By the time of Spindletop, oil had made (and lost) many fortunes, including one megafortune—that of John D. Rockefeller, who founded the Standard Oil Company of Ohio in 1870. Rockefeller took the unique approach of buying up not only wells but sources of everything involved in the refining and distribution of oil, from the oak trees used to make oil barrels to the railroads that shipped the oil to its final users—a process later called vertical integration. With all aspects of the industry under his control, he underbid and devoured competitors until the U.S. government forced his Standard Oil Trust, which controlled his numerous companies, to break up in 1892. Rockefeller transferred his empire's assets to the Standard Oil Company of New Jersey, but in 1911 the Supreme Court broke that company up as well, ruling that it violated the 1890 Sherman Antitrust Act.

The destruction of Rockefeller's empire was not the only change the oil industry was undergoing at the turn of the century. Kerosene lamps were giving way to Thomas A. Edison's 1879 invention, the electric light, so demand for kerosene was falling. At the same time, however, demand for gasoline, a hydrocarbon that boils out of crude oil at a lower temperature than kerosene, was starting to grow because gasoline was the fuel of choice for the internal combustion engines that powered automobiles. These engines were first developed about the time that Drake drilled his Pennsylvania oil well and they grew widespread after Henry Ford began selling his mass-produced Model Ts in 1908. Oil replaced coal as the world's most popular fuel in the mid-20th century, mostly because of its use in transportation.

Forty-five percent of the oil consumed in the United States in 2003 was used as gasoline for motor vehicles.

Petroleum has become vital to almost every other aspect of life in developed countries as well. It is the base for medicines, plastics, diesel and jet fuels, heating oil, and innumerable other industrial and consumer materials and products. Add to this the facts that oil produces more energy per unit volume than coal or gas, is easily transportable, and is relatively cheap for the amount of energy it produces, and it is little wonder that 40 percent of the energy that the United States consumed in 2004 came from oil—more than from any other primary energy source. The whole world uses about 78 million barrels of oil a day. Lita Epstein and the other authors of *The Complete Idiot's Guide to the Politics of Oil* state bluntly, "Without oil, industrial and technological society as we know it would not exist."[3]

Essential as oil seems to be, critics say it presents many threats to health and the environment. Oil drillers and refinery workers, like coal miners, run a high risk of illness, injury, or death. Blowouts from offshore wells and spills from oil tankers such as the famous breakup of the *Exxon Valdez*, which dumped almost 11 million gallons of oil into Alaska's Prince William Sound on March 24, 1989, cause highly publicized environmental disasters that take many years and millions of dollars to clean up. Although the *Valdez* spill was the worst in U.S. history, it has been dwarfed by others, such as the collision of the *Atlantic Empress* and the *Aegean Captain* in the Caribbean in 1979, which spilled 110 million gallons of oil, and the blowout of a Mexican offshore well called Ixtoc I into the Gulf of Mexico, also in 1979, which poured out more than 140 million gallons in a little under 10 months.

Public outcry following major oil spills has produced environmental legislation that aims to minimize spills or, alternatively, keep drilling out of ecologically sensitive areas. Laws were passed to prohibit drilling off many parts of the California coast, for instance, after a blowout of an offshore well near Santa Barbara coated beaches and seabirds with black, gooey crude oil in 1969. The *Exxon Valdez* spill, for its part, resulted in the Oil Pollution Act of 1990, which, among other measures to help prevent and clean up oil spills, orders all oil tankers in U.S. waters to be double hulled by 2015. The strong, lightweight outer hull of a double-hulled ship is meant to take the impact of a grounding or collision, leaving the inner hull and cargo undamaged.

Oil industry representatives claim that the risk of spills has been exaggerated. Most oil that enters the oceans, they say, comes from natural seeps. They point to a 2003 National Research Council report that puts the figure for seeps at 48 percent of total oceanic oil, more than any other single contributor. (A 1995 Smithsonian Institution exhibit, on the other hand, stated that seeps account for only 9 percent of marine oil pollution.) Offshore drilling platforms, industry supporters maintain, are even safer than tankers. Far from being damaged by the platforms, sea life uses them as artificial reefs—a kind of underwater apartment house.

Environmentalists say that spills and blowouts are only part of the pollution from oil extraction and transportation. Rock, mud, and water from oil wells contain toxic pollutants that can contaminate groundwater. Drilling and related activities also produce air pollution, and drilling machinery can blight and disrupt a landscape almost as much as strip mining. Pipelines that transport oil can block animal migration routes, and they sometimes leak. They are also tempting targets for terrorists.

Perhaps the greatest concern of all has been expressed about the effects of burning oil. Burning petroleum or petroleum products such as gasoline give off carbon monoxide, hydrocarbons, nitrogen oxides, particulates, and carbon dioxide. Oil does not produce quite as much pollution as coal, but the amount is still significant. Gasoline formulations and vehicle emissions have been made cleaner in the developed world, for instance, by removing lead from gasoline in the late 1970s, but the small engines that power the scooters, agricultural trucks, and other vehicles popular in developing nations (as well as motor bikes, lawnmowers, leaf blowers, and similar devices in developed ones) emit tremendous pollution.

NATURAL GAS

So-called natural gas is largely methane, a simple molecule made up of four hydrogen atoms surrounding one atom of carbon. It is often found in the same places as oil and is extracted along with it. Gas is normally collected from wells in small-diameter pipelines, processed in the field to remove pollutants and water, compressed to boost pressure, and pushed through large transmission pipelines to storage or marketing centers. From there, local utilities send it to individual buildings through gas mains. Gas can also be chilled to turn it into a liquid (liquified natural gas, or LNG), which can be shipped from country to country in special tankers.

Oil and Gas Journal estimated that the world had 5,505 trillion cubic feet of natural gas reserves at the start of 2003. The Middle East and Russia have the largest reserves. The United States has about 189 trillion cubic feet of reserves, as compared to Russia's 1,680 trillion cubic feet. Texas, Oklahoma, and New Mexico are the top natural gas–producing states. Natural gas accounted for 23 percent of the energy used in the United States in 2003, the same percentage as coal and a little more than half the percentage of oil. The United States imports about 18 percent of its natural gas, almost all from Canada.

Like coal and oil, natural gas was used as a fuel in ancient times in areas where it came out of the ground naturally. Gas was employed fairly extensively for lighting in 19th-century Europe and America. Until an interstate gas pipeline system was completed after World War II, however, natural gas in oil fields was often considered close to useless. It was either burned off in giant flares at wellheads or captured and reinjected into the wells to force more oil out.

Today, people increasingly recognize that natural gas is the most desirable of the fossil fuels because of the high efficiency with which it produces energy and the relatively low quantity of pollutants, including carbon dioxide, that it releases when burned. It is also easy to distribute and use. It heats buildings and runs appliances such as stoves, clothes dryers, and water heaters. Most important, it is a preferred fuel for generating electricity. Combined cycle plants, in which the waste heat from gas-fired turbines is recaptured in boilers and used to heat steam that powers other turbines, are considered the most efficient and least polluting of conventional power plants.

Although natural gas still accounts for a far smaller share of U.S. electric power generation than coal—16 percent, as opposed to coal's 51 percent, in 2003—demand for it is growing as criticisms of coal's environmental effects become stronger. Supporters such as Henry R. Linden, professor of energy and power engineering at the Illinois Institute of Technology, call natural gas "the ideal transition fuel to a sustainable and carbon-emission-free energy system."[4] Environmentalists nonetheless point out that, as a fossil fuel, gas is limited in quantity and produces greenhouse gas emissions; indeed, methane itself is a more powerful greenhouse gas than carbon dioxide.

NUCLEAR POWER

Much as fossil fuels have been criticized, they have never been attacked with the fervor shown by opponents of another conventional energy source, nuclear power. Nuclear or atomic power comes from the energy released when the atomic nucleus of a radioactive element, usually the radioactive form (isotope) of uranium, U-235, is broken apart by the impact of a subatomic particle called a neutron. This process, termed fission, was first demonstrated in Germany in 1938. Other neutrons are among the particles emitted by this atomic splitting, and if the uranium is in a confined mass, these free neutrons can hit additional nuclei and continue the process, producing a chain reaction.

Fission reactions can be either uncontrolled, as in the atomic bombs that destroyed the Japanese cities of Hiroshima and Nagasaki in August 1945, or controlled, as in nuclear plants used to generate electricity. There are several models of such plants, but most have features in common. The core of a typical nuclear power plant contains pellets of concentrated uranium ore sealed inside thousands of long rods called fuel rods. This core is shielded by several thick-walled chambers. Control rods made of a material such as graphite are inserted among the fuel rods to absorb neutrons and slow the reaction down; if a problem arises, more rods can be put in to stop it completely. The reactor also contains a moderator substance that slows the neutrons further and makes the fission process more efficient. Water under very high pressure absorbs the energy released by the reaction, heats up to form steam, and turns turbines to generate electricity.

The uranium that nuclear plants usually use as a fuel is a nonrenewable resource, although supplies are not expected to run out for a century or so. Some of the world's largest uranium deposits are in Australia and central Asia, including Mongolia and the Caspian Sea area, the latter of which has also become famous as a potential major source of oil. The United States also has substantial uranium deposits, almost half of which are in Wyoming.

To conserve uranium supplies, reactors can also use another radioactive element, plutonium, as a fuel. A type called a breeder reactor fissions uranium to produce plutonium, then fissions the plutonium in turn to generate still more energy. Plutonium can also be recovered from the used fuel rods of ordinary reactors, so breeder reactors have the advantage of both minimizing the need to mine new uranium and disposing of a dangerous form of radioactive waste. Using plutonium in nuclear power plants makes many people nervous, however, because, unlike uranium ore, plutonium can be easily made into bombs. For this reason, the United States banned construction of breeder reactors and reactors that reprocess plutonium in the early 1990s. Countries including Russia and Japan, however, do produce plutonium in their nuclear plants.

As of 2004, the United States has 104 working nuclear power plants, which collectively provide about a fifth of the electricity the country uses. This is the largest number of plants in any one nation, and almost a quarter of the world's total, but some other countries are much more dependent on nuclear power than the United States. France, for example, obtained 78 percent of its electricity from nuclear plants in 2002. Nuclear power accounted for only 6.56 percent of the world's primary energy supply in 2001, however, slightly less than the portion that came from hydroelectric power and far less than the amount from fossil fuels.

Richard L. Ottinger and Rebecca Williams said in the Spring 2002 issue of *Environmental Law* that nuclear power was the world's slowest-growing energy source. The first reason for this slow growth is economic. When nuclear energy began to be used commercially in the late 1950s, Atomic Energy Commission chairman Lewis Strauss predicted that in a generation or so, electricity from nuclear plants would be "too cheap to meter."[5] That proved to be anything but the case. Nuclear plants were exceptionally difficult and expensive to build, and many of those ordered in the 1970s were never finished. No new plants were built in the United States after 1979. According to *Forbes* magazine, nuclear power became "a[n economic] disaster of monumental scale."[6]

Environmental criticism, however, has limited nuclear power even more than economics. The greatest concern has been the possibility of radiation release from a plant accident. This fear grew as the environmental movement gathered strength in the 1970s and became overpowering in the wake of an accident at a nuclear power plant in Three Mile Island, Pennsylvania, that began on March 28, 1979. The accident, caused by a combination of equipment failure and human error, leaked a small amount of radioactive material into the surrounding air and water. Although this event was later described as "the nation's worst nuclear accident," supporters of nuclear power point out that it was never proved to have caused physical harm to anyone.[7]

The same could not be said of the accident that occurred at Chernobyl, in the Ukraine, then part of the Soviet Union, on April 25, 1986. There, the reactor overheated and the carbon used as the moderator caught fire, producing a fireball that blew off the reactor lid and released a cloud of radiation that spread all over Europe. Some 30 people were killed, and investigators claimed that radiation from the accident would cause between 50,000 and 70,000 extra deaths from cancer or other diseases within 70 years. Nuclear power advocates say that the type of reactor at Chernobyl was obsolete even at the time and is quite unlike U.S. power plants. They point out that other reactors all over the world have operated for decades since then without any significant accidents.

Some critics fear that radiation release from a nuclear plant might not happen by accident. Especially after the attacks on New York's World Trade Center and the Pentagon on September 11, 2001, many commentators have warned that terrorists might target such a plant. Journalist Kate Tsubata wrote in 2002, however, that "In the estimation of the FBI, the safety measures and defenses of nuclear plants make them the best-protected facilities in the United States, next to military installations."[8]

Other criticisms relate to waste from the plants. Some of this waste, such as depleted fuel rods, is estimated to remain dangerously radioactive for at least 10,000 years—and a single plant can produce 20 tons of such waste each year. At present, radioactive waste is usually stored within the plants themselves, in sealed drums standing in ponds of water. The ponds may soon be replaced with casks of steel and concrete.

The search for a more permanent form of disposal has caused endless debate. In a 1987 amendment to the Nuclear Waste Policy Act of 1982, which set up a national program to dispose of highly radioactive nuclear waste, the government proposed burying such waste at Yucca Mountain, Nevada, 90 miles northwest of Las Vegas. Environmentalists and the Nevada state government strongly objected. For instance, Diane D'Arrigo of the Nuclear Information and Resource Service, an antinuclear group, stated in 2001 that the Yucca Mountain site is near or on 32 earthquake faults and has a "history and prospects of volcanoes and a likelihood of flooding and leakage."[9] Congress finally approved the site in July 2002, but in July 2004 a three-judge panel from a federal appeals court ordered the Environmental Protection Agency to develop a better plan for long-term shielding before proceeding with construction.The facility must also obtain a license from the Nuclear Regulatory Commission before it can be built, and this is expected to be another protracted process.

WATER POWER

Is water power a "conventional" or "alternative" energy source? Is it "green and clean" or a threat to the environment? The answers to these questions depend on who is speaking—and what kind of water power is being talked about.

Water can be made to do work as it runs from a higher to a lower elevation, and humans have used this energy throughout their history. The oldest application of water power was in mills, in which flowing water turned a paddlewheel attached to a drive shaft that rotated one stone above another to grind grain. Eighteenth- and 19th-century textile and other industries also often used water mills for power.

Today, energy from water is usually used to generate electricity. The most common form of water power is hydroelectric power, often shortened to hydropower or simply hydro. Some hydropower installations draw on the natural flow of rivers, but in most large 20th-century hydro plants, a dam is built across a river, creating a reservoir of water behind it. When the gates of the dam are opened, a controlled amount of water falls from the reservoir to a basin below. On its way, it passes over turbines and turns them, producing electricity. The height difference between inflow and outflow is called the head of the dam. The greater the head, the more energy is produced, and the more flooding behind the dam is required.

Hydropower provided 6.62 percent of the electricity used in the world in 2001, making it the most popular primary energy source after the fossil fuels. It accounted for 3 percent of the energy consumed in the United States in 2003. Because hydropower has been used for such a long time (the first hydro installation in the United States was set up in Wisconsin in 1880), it is usually considered to be a conventional energy source. Unlike most other conventional sources, however, it is also renewable. It is by far the most widely used renewable energy resource today, providing almost half of the renewable energy in the United States and 88 percent of that used in the world.

Hydropower is a highly efficient energy source, and once hydro dams and plants are paid for, the energy from them is cheap. Hydro's supporters have also long praised it as environmentally benign. They say it emits essentially no pollutants, and its dams can control natural flooding and provide water for farmland irrigation. The reservoirs behind them often become popular recreation spots.

Especially in recent years, however, environmental groups have painted hydro in less rosy colors. They point out that its dams and reservoirs flood large areas, including some of great natural beauty. Such flooding wipes out ecosystems and forces human residents to find new homes. Unexpected flooding from a burst dam can also cause tremendous damage and loss of life. A significant percentage of water in reservoirs is lost to evaporation as well, and use of water for hydropower ties up supplies that fishers, farmers, and others may need. Hydropower installations change the course of rivers and can block the migration of salmon and other fish.

To avoid some of these problems without giving up the advantages of hydropower entirely, some environmentalists have recommended that plants be smaller and use either the natural flow of rivers, without dams, or dams with lower heads. A dam with a low head and a large volume of water can produce the same amount of electricity as one with a high head and smaller water volume.

Conventional hydropower is not the only way of obtaining electricity from moving water. Several other uses of water, including harnessing the power of waves, tides, and differences in ocean temperature, are classified as alternative energy sources. They are still experimental (probably the only active plant using any of them is one at La Rance, France, which has used tidal power since 1960), and none is likely to ever find widespread use because they work only in places with particular geographical characteristics. They also will probably be very expensive to build and could damage the environment. However, they are completely renewable and produce no greenhouse gases or other pollutants, and supporters think they will eventually prove to be valuable sources of electricity in certain areas.

WIND

Wind energy is really solar energy in disguise. Winds are currents of moving air in the atmosphere, created by the fact that the Sun warms the Earth unevenly. Warm air expands and rises, while cool air contracts and sinks.

The difference in pressure between these two types of air masses (pressure is lower in the warmer air) makes winds blow. About 2 percent of the solar energy that reaches Earth is converted into wind.

Since early times, people around the world have used wind at sea to move ships and on land to turn mills that grind grain and pump water. The Netherlands (Holland) was famous for its windmills, and many farms in the United States used (and some still use) pinwheel-shaped windmills on tall towers to pump water from wells. Today, however, wind power is usually used to generate electricity by turning turbines.

Modern wind turbines, first developed in Denmark in the 1980s, look nothing like the old Dutch windmills. Each turbine is a group of giant, spinning blades (usually either two or three) on top of a tall tower. As Mona Chiang wrote in *Science World*, a magazine for young people, a wind turbine works like an electric fan in reverse: Instead of using electricity to turn the blades to produce moving air, it uses moving air (wind) to turn the blades and provide energy (via a turning shaft and a set of gears) to a generator in the apparatus. Hundreds of turbines are usually grouped together in clusters called wind farms.

Wind turbines today are much larger than those of the 1980s, and their towers are taller. The bigger blades capture more wind, and the higher towers reach up to atmospheric levels where winds are stronger and steadier. Each modern turbine therefore can generate far more electricity than an old one could. Other improvements have made the turbines cheaper, quieter, stronger, more reliable, and able to respond to a greater range of wind speeds.

Germany and Denmark are leaders in European wind power, and among developing countries, India has placed a large stress on harnessing wind. Large wind-power installations also exist in numerous states within the United States, especially California, Texas, Minnesota, and Iowa. The Plains states and the states of the far West are said to have tremendous potential as sources of wind power.

Wind farms are modular, so they are easier to build than many other electric power installations. Wind energy is free as well as renewable and nonpolluting, and the technology of using it is among the best developed of those that harness alternative fuels. Because of these advantages, wind comes closer than any other alternative energy source to being competitive with fossil fuels in price. It therefore is not surprising that, as wind boosters never tire of pointing out, wind power is the fastest growing alternative energy source today. It grew by almost 70 percent between 1997 and 2000, they say.

Nonetheless, wind has some problems as an energy source. Perhaps the biggest one, which it shares with most other alternative fuel sources, is that it is intermittent: The wind does not blow equally strongly all the time. This means that electricity generated by wind is not what is called dispatchable—dependably available to be sent out at any time in amounts proportional to users' demands. The times of day when winds are strongest often are not the times of day when power demand is greatest. Users of wind power therefore must either maintain expensive batteries to store electricity or have a backup generator (usually fossil fuel–powered) to provide electric power when the wind is not blowing. Critics such as Robert L. Hirsch, an energy consultant who has worked for the National Research Council, think that the need for storage or backup will keep wind from ever being truly competitive with fossil fuels. Wind's variability can also cause problems when wind farms are connected to the electric transmission grid.

Even some environmentalists have complained about wind. Wind farms, they say, spoil the aesthetic appeal of the landscapes on which the installations are sited. Early wind turbines were also noisy, although recent designs are much less so. Critics note that wind farms take up tremendous amounts of land that could be used for other purposes, such as housing or recreation. Wind power supporters, however, say that the land occupied by a wind farm can be used simultaneously in other ways, such as for farming or ranching. (Indeed, leasing part of their land for wind farms can provide a welcome income supplement for traditional farmers.) One solution to some of these problems may be to site wind farms offshore. Offshore plants are more costly to set up and maintain than onshore ones, but they encounter stronger and steadier winds, and they are also out of sight.

A final complaint is that birds, especially raptors like hawks and golden eagles, fly into wind turbine blades and are killed. Wind farms at Altamont Pass, California, about 50 miles from San Francisco, have one of the worst kill records—about a thousand birds a year. Makers and supporters of wind turbines point out, however, that these installations were built in the 1980s, and researchers have learned much about birds' interactions with wind turbines since then. The Altamont turbines were inadvertently placed on a major migratory pathway for birds of prey, but builders of new wind farms choose sites where birds are much less common. New tower designs also discourage perching.

Sun (Solar) Power

All energy sources except nuclear, tidal, and geothermal power are, ultimately, solar power—energy from atomic fusion reactions within our star, the Sun. Solar energy also can be harnessed directly to provide heat and electricity.

Some houses and larger buildings get most of their heating and hot water from the Sun. In passive solar heating designs, architectural features essentially make the whole building a solar collector. Many windows face south, but few or none face north. Heavy curtains over the windows keep heat in at night, and awnings reduce the amount of sunlight entering on hot days. Thick floors and walls absorb heat during the day and release it into the building at night.

Active solar heating relies on flat panels called solar collectors. Groups of these panels are mounted on the building's roof, usually on the south side. Each panel is a box containing a metal plate painted black to absorb sunlight. Insulation in the box's back and sides and two sheets of glass above the plate, separated by air or a vacuum, help to concentrate heat inside the box. The panel is connected to pipes filled with water. Even on a cold day, water passing through such a collector can be heated to 140° to 160° F. The heated water can be piped down to the building or a swimming pool, or it can be stored in an insulated tank for later use.

The Sun's heating power is also applied in simple, inexpensive cookers that collect solar energy with a parabolic (curved) mirror about a yard (or meter) across. Solar cookers have proved popular in rural parts of some developing countries, where sunlight is plentiful but firewood scarce and electricity nonexistent. Many analysts think that devices like these will remain the chief use of solar energy for some time to come.

Solar energy can generate electricity, most commonly by means of solar or photovoltaic cells. Bell Laboratories, in Murray Hill, New Jersey, made the first practical solar cells in 1954, and the National Aeronautics and Space Administration (NASA) used them in spacecraft and satellites. Solar cells now exist in several designs, of which probably the most common is a thin sandwich of silicon (the ubiquitous element in sand, glass, and computer chips) containing tiny, deliberately induced amounts of certain other elements such as arsenic and gallium. When sunlight strikes a photovoltaic cell, it triggers a chemical reaction that generates electricity. Each cell produces only a tiny amount of current, but many cells can be packed into compact modules that generate substantial quantities. Photovoltaic cells can be combined with solar panels to provide most of a home's electricity.

Power from modern solar cells costs less than one two-thousandth of an equivalent amount produced by NASA-era cells. Solar-generated electricity is not as cheap as electricity produced by fossil fuels, however, chiefly because the cells are still fairly expensive to manufacture. A few countries, notably Germany and Japan, draw significant power from photovoltaics, but this alternative source of power has proved less popular than wind. The United States has only a few large solar power installations, mainly in California.

Solar power's advantages and disadvantages are similar to those of wind. This energy source produces no greenhouse gases or other pollution (although manufacturing solar cells involves some toxic materials) and is available for free in essentially infinite supply. The cost of maintaining most solar devices is also low, but the cost of building solar power plants and even installing solar collectors is relatively high. Large solar power installations require substantial areas of open land dependably exposed to strong sunlight. Finally, like wind, solar power is intermittent. This makes it an unacceptable sole power source for modern homes, where computers and related devices demand an uninterruptible source of electricity.

BIOMASS

Biomass energy comes from burning plant or animal wastes. For humans it is the most ancient energy source of all. Until the late 19th century, burning wood or animal dung supplied almost all the energy that people used, and in much of the developing world, it still does.

Today, biomass can be burned to generate electricity. Most biomass used in large power installations consists of wood chips, sawdust, "black liquor," and other byproducts of the pulp and paper industries. These wastes are often burned along with coal near the factories that produce them, creating both heat and electricity, a process called cogeneration. Most such factories use the heat and electricity themselves, but some sell extra electricity to utilities. Some cities also burn their solid waste, about 60 percent of which is biomass, to obtain energy. This has the additional advantage of reducing the amount of material that must go into landfills, but the process is by no means pollution-free or cheap.

The second major modern use of biomass is converting it into alternative automobile fuels such as ethanol and methanol. These fuels, whether used alone or combined with gasoline (gasohol), are less polluting than pure gasoline, and they can be used in standard internal combustion engines with little modification. Their use is particularly widespread in Brazil, where all cars and light vehicles have been required since 1979 to use either ethanol or gasohol as fuel. Brazil generates ethanol from bagasse, an agricultural waste left from processing sugar cane. The United States makes ethanol from corn and uses it as a gasoline additive to reduce certain kinds of pollution. Another vehicle fuel that uses biomass is biodiesel, which is made from plant or animal fat.

Most biomass installations use agricultural or other waste, but certain crops can be raised for the sole purpose of being burned. One crop used in Britain is willow trees, which can be fertilized with sewage and harvested about three years after planting. Some woody biomass crops can be grown on land no longer fit for conventional farming. Japan is experimenting with algae as a biomass crop.

"Split wood, not atoms" was a popular saying among environmentalists in the 1970s, but traditional biomass burning can be very harmful to the environment and human health. A need for firewood and charcoal (burned wood used as a fuel) is one cause of the devastating deforestation occurring in much of the world today. Burning wood and animal dung release substantial quantities of particulates and other pollutants into the air. An analysis sponsored by the World Bank estimated in 2001 that pollution produced by these "dirty" energy sources contributes to the deaths of about 2 million people, mostly women and children, each year in India alone.

Modern plants that burn biomass to create electricity produce far less pollution than traditional biomass burning or most uses of fossil fuels. For this reason, even though biomass is not as pollution-free as wind or solar power, it is counted among "green" energy sources. Biomass is considered to be renewable because the crops that produce it are replanted. Burning biomass emits carbon dioxide, but growing the same quantity of new plants absorbs the same amount of the gas, so there is no net increase in atmospheric CO_2 from this source. Like other renewable fuels, biomass is more expensive to use than fossil fuels, and modern biomass applications contribute only a tiny fraction of the energy used in the world and in the United States. It provides a larger share in a few countries, however, including Finland, Sweden, and Denmark.

GEOTHERMAL ENERGY

Heat from the Earth's core, the breakdown of naturally radioactive materials in the crust, and the friction of the crustal plates rubbing against each other make temperatures in the interior of the planet unbelievably high. In places where this heat is brought to the surface as hot water and gases, it has been used to warm buildings and generate electricity. Most such places occur in volcanic areas, where the plates of the Earth's crust are sliding past each other. The first geothermal electric power plant was commissioned in Larderello, northern Italy, in 1913, and the first large-scale use of geothermal power for space heating was in Iceland in 1930.

About three-fourths of the world's harnessed geothermal power is used for electricity generation and the rest for heating. In the limited areas where the appropriate resources exist, making electricity with geothermal power is about as cheap as using fossil fuels. The entire island of Iceland sits on a geothermal reservoir, and that country gets most of its space and water heating, as well as much of its electricity, from this source. About 20 other countries, including New Zealand, France, and Japan, have also built geothermal power plants. The United States has geothermal plants in California, Nevada, Utah, and Hawaii. The largest geothermal electricity plant in the world, in fact, is at The Geysers, California.

At present, all commercial geothermal plants draw on so-called hydrothermal reservoirs, in which steam, hot water, or both lie close to the surface. A few visionaries, however, think that deeper geothermal energy from so-called hot dry rocks may also someday be productively tapped. Unlike hydrothermal reservoirs, this energy theoretically would be available anywhere. A hot dry rock plant would consist of two adjacent shafts reaching deep into the earth, dug much as oil wells are drilled and connected by fractures in the rock. Water would be pumped into one shaft, heated by the rock, and pumped back up through the other shaft to heat steam. When the water had given up its heat, it would be pumped back into the ground to start the cycle again. So far, hot dry rock plants have been created only as demonstration projects.

Geothermal power, at least from reservoirs, is not renewable. Geothermal water, or brine, also contains dissolved minerals, some of which are corrosive or toxic. Removing them and replacing the equipment they damage adds to the cost of maintaining geothermal plants. Overall, however, geothermal power is considered a clean and reliable source of energy, and it could be tremendously valuable if economical ways to tap it on a large scale can be developed.

HYDROGEN

Hydrogen, the most abundant element in the universe, can also be a source of energy. Even groups known for their lack of enthusiasm about alternative energy sources, such as the George W. Bush administration, have expressed excitement about the possibilities of this essentially limitless, nonpolluting, very versatile gas.

Hydrogen is normally harnessed for power in fuel cells, which are something like batteries except that they never need recharging. In a fuel cell, hydrogen combines with oxygen in a chemical reaction that produces water and energy (heat or electricity). The hydrogen can either be provided in pure form or made from other fuels such as methanol by a reformer that is part of the fuel cell. Like photovoltaic cells, individual fuel cells produce relatively little electricity, but they can be stacked in groups to provide a substantial yield. Their chief proposed uses are in power plants to generate electricity or as a substitute for gasoline in motor vehicles.

William Grove, a British physicist, produced electricity by the process used in fuel cells in 1839, but his "gas battery" was simply a laboratory curiosity until 1959, when British engineer Francis Thomas Bacon invented the modern fuel cell. As with solar cells, NASA used fuel cells to provide electricity in spacecraft in the 1960s, but they seemed to have few other potential uses because they were extremely expensive to make. Also as with solar cells, improvements in technology have now greatly reduced fuel cells' cost and increased their efficiency. Several types of fuel cells exist, of which the proton-exchange membrane (PEM) is the most common.

The chief problem with using hydrogen as a fuel is that on Earth it is seldom found in free form, which is required for fuel cells. Instead, it must be obtained from compounds. At present, devices called reformers extract it from natural gas or other hydrocarbons, but ultimately it could be made from water by the process of electrolysis. Because the bonds holding water together are strong, electrolysis requires a considerable input of electricity, which in turn must be made from some fuel. Eventually, proponents of hydrogen power hope to obtain this electricity from renewable sources, but in the short term, economy is likely to dictate that it come from fossil fuels such as natural gas (for stationary cells) or methanol (for fuel cell–powered vehicles). Even if fossil fuels are still part of the process, however, powering cars with hydrogen is potentially valuable because hydrogen fuel cells create no carbon dioxide or other pollutants: the only product of the chemical reaction in a fuel cell is water.

In addition to reducing the cost of fuel cells further and improving their reliability, a number of other technical problems remain to be solved before hydrogen power can see widespread use. PEM fuel cells require platinum, a very scarce and expensive element, which potentially limits their availability and price. Storage of hydrogen can be dangerous because the gas can burn explosively, although supporters say it is no more risky than gasoline. There is also currently no network for distributing hydrogen the way gas stations distribute gasoline. Economically, this is a "chicken and egg" problem: A distribution network and other infrastructure will be built only if there is considerable demand for the product, and people are unlikely to buy hydrogen-powered cars unless these systems are in place.

Nonetheless, the first years of the new century have seen considerable publicity about the possibilities of hydrogen. In his State of the Union Address in January 2003, President Bush promised $1.7 billion in funding for hydrogen fuel cell research and development. All the major auto manufacturers are reported to be developing hydrogen-powered cars, and experimental hydrogen vehicles, mainly buses, are already on the road. Tokyo gains part of its electricity from fuel cell installations as well.

DOMESTIC ENERGY ISSUES

Energy is the lifeblood of a country's economy. Formulating energy policy, therefore, is one of a national government's most important jobs. It is also one of the hardest, because so many factors must be considered: not only the economic needs of producers, consumers, and others involved in energy markets, but also the good and bad social and environmental effects of energy production, distribution, and use. Energy is a major influence on both domestic and foreign policy in virtually all countries today.

In the United States, national energy interests must be balanced against those of state and local governments, which have been regulating the energy industry since its beginning in the late 19th century, and against those of the privately owned companies that control most energy resources and their transmission and distribution. Conflict among these competing groups has shaped (and, many analysts say, often misshaped) the country's domestic energy policy throughout the 20th century.

Basing its authority on the Constitution's Commerce Clause, which has been interpreted to permit Congress to regulate activities within states that affect interstate commerce, the federal government became involved in regulating energy because both economies of scale (complying with one regulation is less expensive than trying to meet half a dozen different ones) and environmental effects force consideration of energy transmission and distribution in regional rather than state terms. Since 1977, energy has had its own Cabinet-level department in the U.S. government, the Department of Energy (DOE).

Several other agencies and departments also play a role in making and carrying out federal energy policy. The most important are the Federal Energy Regulatory Commission (FERC) and the Nuclear Regulatory Commission (NRC), which affect, respectively, the energy industry in general and the nuclear power industry in particular. FERC, which is part of the DOE, regulates interstate shipment of oil, natural gas, and electricity; wholesale (but not retail) electricity sales; accounting and financial conduct of regulated companies; and environmental issues related to some aspects of energy policy. In addition to these, the Department of the Interior (DOI) controls the government-owned lands from which much of the country's domestically produced fossil fuels come, and the Environmental Protection Agency (EPA) enforces federal laws that limit energy activities in order to protect the environment, such as the Clean Air Act. Still other parts of the federal government, such as the Department of Transportation and the Department of Labor, influence particular aspects of energy policy, although they are not major players in forming it.

State public utility commissions and, indirectly, environmental agencies regulate many aspects of energy production and use within state borders. Local governments also affect the energy industry through such actions as regulating land use and granting (or refusing to grant) permits for building new power plants, pipelines, or electricity transmission lines.

Energy has been such a contentious subject that, in recent years, obtaining congressional agreement on a national energy policy has seemed impossible. The last major set of domestic energy policy laws was the Energy Policy Act (EPAct) of 1992. The Bush administration devised a comprehensive energy bill in 2001, but as of mid-2004, it had not persuaded Congress to pass it. Furthermore, environmental and watchdog groups accused the committee that formulated the Bush policy, headed by Vice President Dick Cheney, of being unduly influenced by representatives of the oil and power industries and filed lawsuits to force Cheney to reveal the names of the people whom the committee had consulted. A Supreme Court decision in June 2004, which stressed the importance of protecting the executive branch from interference, made it unlikely (though not impossible) that the groups would obtain this information through the courts, however.

MATCHING ENERGY SUPPLY AND DEMAND

Making sure there is enough energy to meet its citizens' and economy's demands is vital for any country, and many policy concerns in the United States and elsewhere focus on this issue. If energy supplies do not seem likely to match expected demand, the disparity can be removed by increasing supply, reducing demand, or both. Among those who think that energy supplies should be expanded, some favor seeking new domestic sources of fossil fuels, while others prefer increasing the amount of energy that comes from renewable sources, even if this requires government subsidies or other incentives to overcome these fuels' economic disadvantages.

In an emergency, the United States could temporarily expand its energy supply by "drawing down" the Strategic Petroleum Reserve (SPR). The Energy Policy and Conservation Act established this stockpile in 1975, shortly after the 1973–74 Arab oil embargo, to prevent the country from ever being cut so short of petroleum supplies again. The SPR can hold up to 700 million barrels of oil, stored in underground caverns along the coasts of Texas and Louisiana. In October 2004, it held about 670 million barrels. The SPR is the largest emergency oil stockpile in the world.

Both the desirable size of the SPR and the circumstances under which oil from it should be released into the market have been debated. A large SPR provides a more effective buffer than a small one, but it also reduces the amount of oil available for current use. The Bush administration ordered the DOE to fill the SPR to capacity in November 2001, following the September 11 terrorist attacks, but in March 2004, more than 50 members of the House of Representatives asked the president to stop the filling temporarily so that more oil could remain in the market to help lower gasoline prices, which were reaching record levels. In response, U.S. energy secretary Spencer Abraham insisted that filling the SPR had "negligible" effects on gas prices.

The act that established the SPR said the president could decide to draw down, or release, SPR supplies in the event of "a severe energy supply interruption," which was defined as including both cutoffs of imports and shortages that "may cause major adverse impacts on national safety or the national economy."[10] Politicians and economists disagree on whether price spikes are covered by the latter definition and, if so, at what price level SPR supplies should be released. President Bush said in his 2000 election campaign that the SPR is "meant for a sudden disruption of our energy supply or for war," not to reduce high domestic oil or gasoline prices, but an Energy Information Administration fact sheet revised in April 2004 included oil price spikes in the definition of supply interruptions that could harm the economy enough to trigger release from the SPR.[11] Commentators are also divided on whether such release, if it did occur, would effectively lower oil prices.

In any case, the SPR is clearly only a short-term solution to energy shortages. In order to meet the demands of a growing population and an increasingly energy-intensive economy, Bush administration and petroleum industry spokespeople, among others, say that the United States must increase its domestic supply of oil, natural gas, and coal. To achieve that goal, they would like to see both incentives for new fossil fuel exploration and removal of environmental protection rulings that place many of the most promising areas off-limits to would-be extractors.

The most controversial proposed domestic drilling site is the Alaska National Wildlife Refuge (ANWR), a 19-million-acre tract just east of the North Slope, a prolific oil-producing area. Environmentalists call the refuge "America's Serengeti" and say it is an essential habitat for mating and calving caribou, grizzly bears, musk oxen, and geese.[12] Oil industry representatives, on the other hand, point out that they want to drill in only a tiny part of the refuge, some 2,000 acres, and claim that few animals live in this area during the winter, when most drilling would occur. U.S. interior secretary Gale Norton described the site as "flat nothingness."[13]

Oil supporters and environmentalists also disagree on how much damage drilling would do to the site. The oil companies say that new technology makes drilling much less destructive to the environment than it used to be, with a drill site in 1995 covering only 8.7 acres, as compared to 65 acres for a comparable site in 1970. The ability to drill multiple wells with the same rig also cuts down on the number of rigs. They propose to make many roads and other constructions out of ice, which would simply melt away in the spring. Environmentalists, for their part, point to reports issued by the General Accounting Office and the National Academy of Sciences in 2003, which say that effects of gas and oil extraction in national wildlife refuges and on the North Slope have been variable. Little harm has been documented in some cases, but significant damage has occurred in others. The National Academy reported that North Slope drilling produced changes in bowhead whale migration and caribou reproductive success, for example.

The opposing groups argue about how much oil ANWR could provide, too. The U.S. Geological Survey (USGS) has estimated that this source could provide between 5.7 and 16 billion recoverable barrels of oil, enough to meet the country's petroleum demands for between six months and several years. However, the Alaska Coalition, which opposes drilling in ANWR, claims that a more realistic assessment is only 3.2 billion barrels, less than six months' worth of consumption at present rates. Even the USGS has said that the lower end of its estimate is more likely to be correct than the higher one.

Environmentalists interpret these figures to mean that a priceless habitat would be destroyed for little benefit. They also point out that, even if drilling started immediately, no oil from ANWR would reach the market for five to 10 years. Jerry Taylor, director of natural resource studies at the libertarian Cato Institute, wrote in 2001 that even then, the increased supply might have little effect on the world price of oil because Middle Eastern countries such as Saudi Arabia can produce oil so much more cheaply than the United States can. Conservationist Amory Lovins also notes that oil from any site in Alaska currently must be shipped to the lower 48 states through the Trans-Alaska Pipeline System, which he calls a vulnerable target for terrorists.

If domestic energy supplies are to be increased, environmentalists say, the expansion should come as much as possible from alternative energy sources because these cause so much less pollution than fossil fuels. Most such groups therefore support government subsidies and other financial incentives for development and use of alternative energy. Some actual or proposed incentives are federal, while others are administered by the states.

One form of incentive is the renewable portfolio standard (RPS), in which a public utility commission requires utilities to obtain a certain percentage (usually about 6 percent) of their energy from renewable sources. The utilities can generate electricity from such sources themselves, buy from other generators who use renewables, or purchase "renewable energy credits" from other utilities that use more than their own share of renewables. Eighteen states had adopted renewable portfolio standards as of 2004. Great Britain launched a combination renewable portfolio standard and trading program called the Renewables Obligation in summer 2000, and Denmark and Germany have similar programs. Criticizing this type of incentive, Jerry Taylor and Peter VanDoren of the Cato Institute, writing in *Power Economics* in March 2002, say that "moderate RPS programmes accomplish little and aggressive RPS programmes would prove quite expensive."[14] They claim that such programs will not help the environment much because because most of the "renewable" energy will be biomass cofired with coal.

Another incentive is "green pricing," in which utilities give consumers the option of paying a premium price for electricity guaranteed to have been generated by renewable resources. In 2002, utilities in almost 30 states offered green pricing. Consumers have often told pollsters that they support such programs, but experience has shown that only a tiny percentage of people given a chance to sign up for them actually do so.

Still other incentives include net metering, in which consumers can generate their own energy from renewable sources (probably solar power for residential users and cogeneration for industries) and sell any extra power back to their local utility, in effect making their electric meters "run backwards." About 35 states offer this option. Alternatively, some states levy a small fee called a system benefit charge on consumers, in amounts proportional to their energy consumption, and use the money to fund programs for developing renewable-energy technologies or improvements in energy efficiency. As of 1999, 15 states had these. Others offer rebates or tax reductions for such things as installation of solar panels or purchase of gas-electric hybrid or ethanol-using cars. Some states, such as Colorado, allow businesses that violate air-quality standards to reduce their penalties by investing in renewable energy.

Those who favor subsidies and other incentives for renewable energy sources say that such support spurs invention and investment in these relatively new technologies and overcomes some of the economic disadvantages caused by the high cost of building plants that use them. They also assert that fossil fuel industries receive subsidies, often hidden ones, so providing similar support for renewables simply "levels the playing field." In a 2001 book called *Perverse Subsidies,* Norman Myers and Jennifer Kent claimed that fossil fuels and nuclear power obtained some $21 billion in subsidies in the United States, at least 10 times the amount given to renewables.

Critics of subsidies for renewables, however, say that these government gifts artificially support technologies that are not presently, and may never be, economically competitive on their own. Jerry Taylor and Peter VanDoren of the Cato Institute say that a better way to help fossil fuels and renewable sources compete fairly in the market would be to discontinue fossil fuel subsidies. Similarly, Joel Darmstadter, a senior fellow of Resources for the Future, warns that underpricing renewable energy through subsidies simply worsens the encouragement of overconsumption that underpricing of fossil fuels creates.

Environmental groups and some economists claim that the best way to end the gap between energy supply and demand is not to increase supply but, rather, to reduce demand through energy conservation. Critics have equated conservation with living a spartan life—"freezing in the dark"—but most modern conservationists stress increasing efficiency rather than giving up luxuries. "Because everything is more efficient, [a] new home can be larger and more comfortable without requiring additional power," Gregg Easterbrook, senior editor of the *New Republic*, wrote in that magazine in 2001.[15] He pointed out that energy consumption per dollar of gross domestic product dropped in the United States between about 1975 and 1995, yet Americans' standard of living rose during the same period because devices such as refrigerators became more efficient.

Of all the areas of energy use in which to improve efficiency, many analysts say, transportation is the most important. Gasoline for motor vehicles accounted for 45 percent of the oil used in the United States in 2003, according to the EIA. Increasing automobile fuel mileage, therefore, could have a substantial effect in reducing demand for oil as well as the quantity of pollutants released into the air.

In 1975, as one of many energy conservation and efficiency measures included in the Energy Policy and Conservation Act, Congress set up the Corporate Average Fuel Economy (CAFE) program, which established requirements for the minimum miles per gallon that cars and light trucks should achieve. Vehicle manufacturers were required to average at least this minimum over all the models in each category that they produced; in other words, some models could have less than the minimum efficiency, as long as an equal number of other vehicles had more than the minimum. The standards, 27.5 miles per gallon for passenger cars and 19.5 miles per gallon for light trucks, were to be achieved by 1985.

The CAFE standard for trucks was changed to 20.7 mpg in 1996, but the requirement for automobiles is still the same. Critics claim that neither standard has been met for years. *Business Week Online* reported that actual average mileage in 2000 was only 24.2 mpg for cars and 17.3 mpg for trucks. Worse still, the sport utility vehicles (SUVs) that became so popular during the 1990s and early 2000s are considered to be light trucks, so they need to meet only the less stringent truck standard. The largest SUVs are classified as heavy trucks and therefore do not have to meet any mileage standards at all.

Many environmentalists and conservationists think that the CAFE standards badly need to be revised upward—and enforced. The auto industry and others who think the standards should be left alone, however, claim that raising them would require making cars smaller, lighter in weight, and therefore less safe. Supporters of increased efficiency say that this goal could be achieved by means other than reducing weight and that, in any case, a reduction in weight would not necessarily make cars less safe.

Those who oppose revision of the CAFE standards also insist that substantial increases in mileage would be difficult, if not impossible, to achieve with present technology. A 2002 National Academy of Sciences study and numerous others, however, say that recent improvements in engine and drive-train efficiency such as variable valve timing, direct fuel injection, and continuously variable transmissions could be applied to increasing fuel efficiency rather than being used to boost horsepower and acceleration, as they are at present. Mileage could also be improved by building car bodies of lightweight but strong carbon-fiber composite materials, they claim. Even such simple measures as driving at slower speeds and keeping tires properly inflated can improve efficiency. In April 2003, the National Highway Traffic Safety Administration agreed to revise the CAFE standard for light trucks upward by 1.5 mpg by 2007.

Meanwhile, in search of greater efficiency and less pollution, some automakers are beginning to look beyond gasoline-powered cars. "Hybrid" cars such as the Toyota Prius, which have both a gasoline engine and an electric motor, have been on sale in the United States since about 2002 and, although they still account for only a tiny share of the auto market, are growing more popular. They can travel between 45 and 60 miles on a gallon of gas. American manufacturers are expected to produce their own hybrids soon, including hybrid SUVs. Many are also developing hydrogen-powered vehicles, but even General Motors, considered a strong supporter of this technology, does not expect to see hydrogen cars on the road before 2010.

Another good place for families to look for energy savings, conservation supporters say, is in their homes. Better insulation, double- or triple-glazed windows, and plugging of air leaks, for example, can reduce the need for artificial heating and cooling. Use of fluorescent rather than incandescent bulbs and purchase of energy-efficient refrigerators and air conditioners can cut down demand for electricity. Alex Wilson, editor of *Environmental Building News*, said in mid-2002 that these kinds of changes could save 30 to 50 percent of many homes' energy use with very little cost. Businesses can also find many ways to reduce their energy use with little expense or effort, says conservationist Amory Lovins of the Rocky Mountain Institute.

In spite of educational campaigns to help people reduce their energy use, however, studies have shown that businesses and households still usually make energy-related purchase decisions, such as choosing a new refrigerator or a new car, on the basis of initial price rather than energy efficiency. Gregg Easterbrook of the *New Republic* points out that excessively high mandated efficiency standards can even backfire by adding so much to the cost of new appliances that they make people decide to keep their old ones.

Where incentives for energy conservation fail, some analysts say, disincentives for waste may succeed. Washington University (St. Louis) professor Murray L. Weidenbaum, for instance, recommends removing caps on energy prices so that consumers will become aware of the true cost of the power they use. Alternatively, taxes can be placed on general energy use, on specific fuels such as oil or gasoline, or on carbon emissions. However, several commentators have called the idea of such taxes "a political nonstarter." One objection to energy taxes is that they would disproportionately affect poor people.

In any case, some critics question how effective conservation and increased efficiency can be at reducing energy demand, even if they are achieved to some extent. "The idea that conservation can have a huge impact on energy demand . . . is unrealistic and even dangerous," writes Michael Economides of the University of Houston.[16] Margaret Fels, a Princeton University energy researcher, has claimed that energy savings from conservation tend to amount to only half of what is expected. Peter Huber, a senior fellow at the Manhattan Institute, claims that improving efficiency actually increases energy use because it makes energy-consuming activities cheaper. He calls this effect the "efficiency paradox."

Many commentators, such as *U.S. News & World Report* columnist Mortimer B. Zuckerman, say that both conservation and new sources of energy are necessary to ensure that domestic energy supplies will match present and future demand. "We cannot conserve our way out of the problem. Nor can we drill our way out of it," he wrote in early 2002. "We need production and conservation."[17]

POLLUTION

Another vital issue in any modern country's energy policy is minimization of the environmental damage caused by energy use. In recent years, much stress has been laid on the air pollution produced by fossil fuel burning, especially in motor vehicles and electric power plants. Reduction of air pollution in the United States is governed by the Clean Air Act (CAA), which states implement under the supervision of the Environmental Protection Agency (EPA).

More than half the air pollution in the United States comes from motor vehicles. Among other things, the Clean Air Act gives the EPA authority to specify the composition of vehicle fuels in order to reduce pollution. This provision was used to phase out lead in gasoline beginning in 1975, for instance. Most recently, controversy has centered on oxygenating (oxygen-containing) compounds that the 1990 revision of the CAA mandated adding to gasoline to reduce carbon monoxide pollution during the winter in some parts of the country. Most refineries chose a substance called methyl tertiary butyl ether (MTBE) as their oxygenating compound.

Around the end of the 1990s, however, research showed that MTBE often polluted groundwater and drinking water, possibly harming health. Individuals and communities began suing MTBE manufacturers and gasoline makers to recover damages and cleanup costs. The 2003 version of the Bush administration's national energy policy bill contained a provision that would have protected companies from such suits. This provision was widely held to have been more responsible than any other for the Senate's rejection of the bill.

After MTBE began to be phased out, many refineries added ethanol to gasoline as an oxygenating compound instead. Critics such as Peter Navarro, a business professor at the University of California, Irvine, have complained that this compound does little to improve air quality, raises the price of gasoline because it is usually in short supply, and requires expensive facilities for storage, refining, and distribution. Supporters such as Brooke Coleman of the Renewable Energy Action Project, however, praise ethanol's ability to reduce carbon dioxide as well as carbon monoxide emissions.

Fossil-fuel-burning, especially coal-burning, electric power plants have come in for almost as much criticism regarding air pollution as motor vehicles. Electricity generation produces about 75 percent of the sulfur dioxide emissions and 33 percent of carbon dioxide emissions in the United States.

In 1970, amendments to the Clean Air Act stated that new power plants of any kind, as well as old ones that were substantially remodeled, would be required to meet strict "new source review" standards for emission of certain pollutants. Plants already running in 1970 and not remodeled, however, did not have to meet these standards. A 2003 article in *Nation* estimated that about three-fourths of pollutant emissions from U.S. power plants come from such older plants, which usually burn coal.

Both utilities and environmentalists are critical of these regulations. They say that, because power companies in many states can no longer be sure of recouping the costs of building or improving their plants, the rules make it more cost effective for the companies to keep using their oldest and dirtiest plants than to remodel or replace them with improved technology. The two groups differ on how the rules should be changed, however. Environmentalists would like to see the exemption for older plants removed, whereas utilities would like, instead, to remove the requirement for remodeled plants to meet new source review standards and even to weaken the standards for new plants. In late 2003, the Bush administration ruled that older plants could be remodeled without having to meet the new source review standards.

One program set up to control pollution through the Clean Air Act, the acid rain program established in the act's 1990 revision, has drawn praise as a model of a successful cap-and-trade program, the kind of program proposed to control worldwide emission of greenhouse gases. The program set a limit, or cap, for total national sulfur dioxide (SO_2) emissions and gave each plant a certain number of "emission credits" that specified how much

SO_2 it individually could produce. If it emitted less than the amount assigned, it could sell its unused credits to other companies that could not, or chose not to, reduce their emissions enough to meet the law's standards. The effect was that the aggregate amount of pollution was significantly reduced at the least possible cost, even though some plants continued to pollute above the standard levels. The program has been lauded for using market mechanisms rather than regulatory standards to achieve its ends.

In addition to traditional pollutants such as sulfur dioxide, electric power plants that burn fossil fuels, especially coal, produce considerable carbon dioxide. The United States, in fact, produces more carbon dioxide and other greenhouse gases than any other country—about a quarter of the world's total. There is growing debate about whether the Clean Air Act should be revised to include CO_2 as a pollutant and set standards for its reduction. Environmentalists favor this move, but most energy utilities and fossil fuel companies oppose it because they claim there is no affordable way to remove the gas from power plant emissions.

Some other countries have already taken steps to limit their domestic production of carbon dioxide and other greenhouse gases. In 2000, for example, Great Britain placed a tax called the Climate Change Levy on energy use by manufacturers and energy suppliers. Money gained from the tax is used to help the country meet its CO_2 reduction goals. Businesses can reduce their share of the tax by lowering either their energy use (relative to production levels) or their greenhouse gas emissions or by using more energy from renewable sources (except hydropower). Britain also has a voluntary Emissions Trading Scheme for CO_2, similar to the U.S. cap-and-trade program for sulfur dioxide.

Some people favor finding ways to keep carbon dioxide out of the atmosphere rather than preventing its emission. Planting forests, for instance, creates "carbon sinks" that can absorb the gas. Another possibility is sequestration, in which CO_2 is injected into old oil or gas wells or coal beds or buried in the deep sea or underground caverns. Alternatively, the gas could be bound into carbonate minerals and made into construction materials. Techniques for carbon sequestration are still in their infancy, and their effectiveness and possible environmental drawbacks are unknown.

ENERGY INDUSTRY DEREGULATION AND RESTRUCTURING

The electricity and natural gas industries, like the oil industry, began in the late 19th and early 20th centuries as essentially unregulated private firms. Then, as now, both industries included several stages: generation, in which gas is taken from the earth and electricity is created from other forms of energy; transmission, in which large amounts of gas or electricity are sent over long distances from generation points to the general areas where they will be used; and distribution, in which they are transported (often in somewhat modified form) to the industries and businesses that are their final users.

The electricity industry grew out of Thomas Alva Edison's invention of the electric light in 1879. Edison opened the first central electricity generating station in the United States, the Pearl Street Station in New York, in 1882. This and other early generating stations produced electricity in the form of direct current (DC)—current that always flows in the same direction. Because direct current could not be transmitted more than about three miles (five kilometers) without requiring large, expensive copper cables, generating stations had to be close to their customers; indeed, many large buildings, such as stores, produced their own electricity.

In the 1890s, however, U.S. inventor and manufacturer George Westinghouse began promoting the use of alternating current (AC), in which the current flows back and forth in a wire. Unlike direct current, alternating current can be "stepped up" to high voltage with a device called a transformer. It can then be sent considerable distances through relatively thin cables, because high-voltage electricity is transmitted more efficiently (with less loss of energy) than low-voltage current.

Samuel Insull, a former assistant to Edison, realized that this advantage of AC could be used to transmit electricity efficiently and cheaply. Electricity could be generated as direct current at a small number of central power plants, changed to AC by devices called converters and stepped up to high voltage, and sent all over a city through a network of wires. The network would connect to substations in different parts of the city, where other transformers and converters would change the electricity back into low-voltage DC for use by nearby homes and businesses. By 1900, this became the standard way electricity was transmitted. Transmission was soon extended to states and then to multistate regions.

As president of the Commonwealth Edison Company, which transmitted and distributed electricity, Insull also came to recognize, as John D. Rockefeller had with oil, that money was to be made by establishing and controlling all aspects of his business. Just as Rockefeller had taken over not only oil wells but railroads and other industries needed to package and distribute the oil, Insull bought up not only electric utilities in several states but construction companies and other businesses that supplied the utilities. This "vertical integration" made Insull and those who followed in his footsteps rich, just as it had done for Rockefeller, because it meant that they had a monopoly, or complete control over an industry, in their different geographical areas and therefore could set virtually any price they liked for their product. The same thing was happening with the natural gas industry, where some companies bought not only gas wells but the networks of pipes that were growing up to transmit the gas from wellheads to customers.

Regulation of Energy Utilities

Even in those freewheeling early days, the energy industry had to contend with some government regulation. According to common law that traced back to England in the Middle Ages, governments have the right to regulate a public utility, defined as a private business that provides an essential service to the public and is granted the right to be the only one offering that type of service in a given territory. In return for this monopoly privilege, the government can require that the utility serve all customers and charge prices just high enough to cover its costs plus a "reasonable" rate of return to its investors. The U.S. Supreme Court affirmed this governmental right in an 1876 case, *Munn v. Illinois.* Governments classified gas and electric companies as public utilities, both because the companies' services were increasingly seen as indispensable and because they were perceived to be natural monopolies, types of business in which a single company can provide a good or service more efficiently and cheaply than two or more competing firms can. Building competing systems of gas pipelines or electric transmission wires in the same area would have been very wasteful, for instance.

Local governments regulated gas and electricity utilities from their beginnings. In the early 20th century, after concluding that local regulation was variable, ineffective, and sometimes corrupt, state governments added a layer of oversight in the form of what came to be called public utility commissions or public service commissions. Every state today has such a commission. Most regulation of utilities is still done at the state level.

The federal government took its first step into energy regulation in 1920, when Congress passed the Federal Water Power Act and the Mineral Leasing Act. The first of these laws established an agency called the Federal Power Commission to set up and regulate large hydroelectric projects on federal lands. The second created rules that governed leasing of federal lands for oil and gas extraction.

By the early 1930s, it was clear that energy needed stronger and more comprehensive federal control. For one thing, many electricity and gas utilities had come to be owned by holding companies—businesses formed for the sole purpose of owning and controlling other businesses. The largest holding companies owned groups of utilities and transmission networks that stretched across multiple states, much as John D. Rockefeller's trust had done with oil-related companies in the 1880s. State regulation therefore could not touch them.

When holding companies' abuses of power became obvious, the public demanded that the federal government take action. The result was the Public Utility Holding Company Act (PUHCA), passed in 1935 as part of a larger package called the Public Utility Act. PUHCA required holding companies to limit themselves to specific geographic areas and confine their business in other ways. A second part of the Public Utility Act, the Federal Power Act, restructured the Federal Power Commission and expanded its duties to include regulation of interstate transmission and wholesale sales of electricity. Natural gas transmission and wholesale sales were added to the list of regulated activities in 1938.

Numerous power plants were built between the 1930s and the 1970s, and nationwide networks of pipes and power lines were constructed to bring gas and electricity to cities, towns, and eventually rural areas. Cheap energy helped the American economy expand and the standard of living rise after World War II, so demand for power increased steadily. For the most part, consumers enjoyed low prices and reliable service, utilities received guaranteed paybacks on their investments and some profit, and everyone was relatively happy.

This situation changed in the 1970s, when the shock of the Arab oil embargo, the rise of the environmental movement (which warned of the pollution produced by burning fossil fuels and the risk of radiation leaks from nuclear plants), and skyrocketing oil prices produced a new emphasis on conservation and efficiency. The result was a sharp drop in energy demand. Utilities suddenly discovered that they had, or had ordered, more power plants than they needed. Consumer groups complained that the utilities had become inefficient and wasteful because the existing regulatory system guaranteed that they would recoup their costs. Economists also began pointing out that, while electricity and gas transmission and distribution might be natural monopolies and thus in need of regulation, generation really was not.

Congress came to believe that "functionally unbundling" the vertically integrated utilities to separate generation from the other functions and allowing open competition in generation would improve utilities' performance, bring new and less polluting technologies into the energy mix, and reduce prices for consumers. With laws passed in 1978 as part of the National Energy Act, therefore, it started a complex process that has come to be known as deregulation, although many commentators feel that *restructuring* is a better term because considerable regulation of utilities still exists.

The natural gas industry suffered—and survived—its deregulatory growing pains before electricity. In a key 1954 court case, *Phillips Petroleum v. Wisconsin*, the Supreme Court had ruled that the Federal Power Commission had the authority to regulate wellhead as well as transmission prices for natural gas shipped across state lines. This resulted in an extremely confusing system that several scholars in the field have blamed for gas shortages in the late 1970s because it placed government controls on an industry (gas production) that was not a natural monopoly and therefore did not really need regulation.

To correct this situation, Congress passed the Natural Gas Policy Act (NGPA) in 1978. The NGPA aimed the gas market toward deregulation, but it expected the move to take many years as gas prices and supplies changed slowly. Instead, the combination of high prices and conservation produced a fairly rapid drop in demand, resulting in an increase in supply and a drop in wholesale prices. Pipeline companies and local distributors, however, were locked into long-term "take or pay" contracts that required them to pay their suppliers a certain minimum amount for gas whether they actually took the gas or not, guaranteeing the suppliers a minimum cash flow in return for their guaranteeing a gas supply to the purchasers. The prices in these contracts remained at the old, high level, so the companies—and, therefore, consumers—still had to pay high prices for gas, even though it now existed in surplus.

Between 1985 and 1992, however, the Federal Energy Regulatory Commission, which had replaced the Federal Power Commission when the Department of Energy was created in 1977, was able to solve these problems. FERC Order 436, issued in 1985, and Order 451, which appeared the following year, pressured pipeline companies to carry gas from distributors they did not necessarily own and made pricing between producers and pipelines more flexible. This meant that distributors and industrial consumers had access to many producers rather than being forced to accept the producers the pipelines chose, and producers and pipelines had to lower their prices in order to compete. Consumer prices of gas fell, demand rose, and the surplus began to disappear.

Natural gas deregulation (except for transmission pipelines, which remain regulated as a natural monopoly, and distribution companies, which are regulated by state public utility commissions) was completed with the Natural Gas Wellhead Decontrol Act, passed in 1989, which removed all wellhead price caps as of January 1, 1993, and FERC Order 636, issued in April 1992, which made it mandatory for companies selling natural gas to functionally unbundle their gas-producing divisions from their pipeline and storage divisions and to make their interstate pipelines available to all producers on the same terms. Unbundling separated parts of the industry still perceived as natural monopolies, in this case gas pipelines, from parts no longer seen that way, here gas production, and required the monopoly part (transmission and storage facilities) to be open to all competitors. After this, distribution companies and large industrial customers bought their gas directly from suppliers and contracted separately with pipeline companies to transport it, rather than buying it from the pipeline companies as before.

The increased competition brought about by these changes encouraged technical improvements that cut the cost of producing, storing, and transporting gas. It also helped to integrate the patchwork of gas transmission systems into a seamless network that covers all of the United States and Canada. Therefore, many analysts feel that deregulation in the natural gas industry was basically successful. However, only large consumers benefited from the market's new openness; small ones were still captive to the pipeline and distribution monopolies.

Electricity Industry Restructuring

The restructuring path for the electricity industry has been considerably bumpier than that for natural gas. Part of the reason arises from certain physical characteristics of electricity that gas does not share. First, unlike gas, electricity cannot be stored in large amounts (batteries can store only relatively small quantities); it must be transmitted and used right after it is produced. This means that increased demand must be matched immediately by increased generation. It also means that proper management of the relationship between supply and demand from moment to moment over large (multistate) areas is crucial. This difficult task became even harder under deregulation because many different companies began using the same transmission network.

Second, electricity always flows along the path of least resistance and tries to distribute itself evenly throughout a network. This means that parts of transmission networks can overheat or suffer other problems if too much or too little electricity flows through them, just as a fuse in a house "blows" or a circuit breaker trips if too much current passes through it. A related characteristic is that electrons are all the same, regardless of where they come from. Gas pipeline owners can use valves to separate one company's gas from another's in their pipes, but no equivalent of valves exists in electricity transmission. These two features make pricing difficult under deregulated competition because current may not follow anything like a straight path from generator to consumer.

Several economic issues have also made electricity deregulation painful. One is the question of so-called stranded costs: expenses incurred in building or improving generating plants (which require a great deal of capital investment) or power lines that a utility was guaranteed to recover under traditional rate regulation but may or may not recover under competition. The most striking example of stranded costs in electricity deregulation was the cost of nuclear power plants that were built but never used, both because their generating capacity was not needed and because public opinion turned so strongly against this source of energy. Other stranded costs were expenses incurred in complying with environmental or other regulations and the cost of fulfilling take-or-pay contracts negotiated when prices were higher or demand greater.

Opinions differ on what percentage of stranded costs a utility should be entitled to recover. Utilities say that if they cannot be sure of making back most or all of their stranded costs, they will be unable to attract the investment capital they need to build new plants and lines or repair old ones. Critics of the industry, however, insist that allowing too much stranded cost recovery encourages utilities to make unwise or inefficient decisions, such as ordering more power plants than they need.

People also disagree about who should pay stranded costs. The money could come from utility ratepayers in the form of charges added to electricity bills, from state taxpayers, from new entrants into the market (to "level the playing field" because they do not have equivalent costs), or from customers who formerly bought power from a utility but, under deregulation, are no longer obliged to do so. None of these groups is eager to pay what consumer organizations often call a "bailout."

A second economic problem concerns price caps and determination of rates for electricity users. State governments and, sometimes, FERC have employed these tools to protect consumers, especially residential consumers, from the worst shocks of fluctuating power prices. If price caps force a utility to sell power at or below its cost, however, the utility may be unable to pay its own creditors and will have to declare bankruptcy. Critics of price and rate caps also say that such caps encourage overconsumption by keeping consumers unaware of the real cost of electricity.

Electricity deregulation and restructuring, like that of natural gas, began in 1978 as part of President Jimmy Carter's National Energy Act. The key law in this case was the Public Utilities Regulatory Policies Act (PURPA), which authorized FERC to require that utilities buy power from nonutility "qualifying facilities" at avoided cost, that is, the price a utility would have paid to generate or buy the power itself. PURPA defined qualifying facilities as small generators (less than 80 megawatts) that used renewable power sources or businesses that cogenerated electricity as a byproduct of industrial processes. This law, which took effect in 1980, caused a mini boom in the technologies used by qualifying facilities, especially cogeneration. It also confirmed the idea that competition was practical in electricity generation, as many nonutility generators showed themselves both able and eager to enter the market and provide electricity more cheaply than traditional, regulated utilities.

PURPA also authorized FERC to order companies that owned transmission networks to carry electricity generated by other companies, a process called wheeling. (A 1973 Supreme Court case, *Otter Tail Power Co. v. United States*, had already indicated that courts could order wheeling.) This provision applied only to wholesale electricity moved between states and was not much used. However, the Energy Policy Act of 1992 (EPAct) clarified and broadened FERC's wheeling authority. It also exempted certain nonutility wholesale electricity generators from restrictions imposed by the 1930s law, PUHCA, and, indeed, from most regulation. This feature of EPAct helped to establish an open market in wholesale electricity. One of the companies that qualified as a so-called exempt wholesale generator was a former gas pipeline firm called Enron.

In 1996, as an implementation of part of EPAct, FERC issued Order 888, which functionally unbundled electric utilities, much as had been done with gas utilities three years earlier, by ordering them to transmit electricity from all third-party generators at the same price they would charge generators that they owned. This requirement of open access to the transmission system became an essential feature of electricity deregulation. Perhaps to soften the blow of these changes, Order 888 also stated that electric utilities could recover most of their stranded costs.

The new rules of the electricity game lured new players to join traditional utilities in the power market. Some were individual nonutility generators. Others were "powerhouse" companies such as California's Calpine, which built, bought, and operated numerous power plants nationally or internationally. Still others were power marketers, also called energy merchants, and power brokers, which owned no electricity facilities themselves but brought together generators, utilities, and others and helped them buy and sell power to one another. In that era of increased competition, where far more buy-sell transactions took place than under the old system, these companies performed a useful function and earned a substantial profit from doing so. Enron Corporation was one of the largest, and eventually most notorious, of the power marketers. FERC regulated power marketers as public utilities but had no control over brokers.

The California Energy Crisis of 2000–2001

In spite of these federal law changes, state governments still had the right to determine how—or whether—they regulated electricity sales within their borders. Some states began deregulating as early as 1996, while others approached the process more hesitantly or did not take it up at all. California was probably the first to try deregulation, using a set of rules that were signed into state law in 1996 and went into effect in 1998. Two years after that, the state had become everyone's favorite illustration of the many ways that electricity deregulation could go wrong.

California set up a complex system. It required the state's three utilities to divest themselves of at least half of their fossil fuel–powered generating plants, which meant that by late 2000, the utilities were producing only 28 percent of the state's electricity and had to buy the rest. It established a short-term market for buying and selling electricity called the Power Exchange and required utilities to trade there rather than establishing their usual long-term contracts with suppliers. It allowed wholesale power prices to change with the market but capped retail prices at 10 percent below current market rates. It gave control of the transmission grid to an independent nonprofit organization called an Independent System Operator.

The flaws in this system began to reveal themselves in May 2000. The weather was unusually warm, and electricity demand rose as Californians turned on their air conditioners. Increases in the state's population and the electricity-intensive computer industry of Silicon Valley also contributed to the growing demand. For reasons that are still being debated, the supply did not appear to be up to the challenge. Wholesale prices skyrocketed in the electricity "spot market," which changed daily, and the state utilities had to pay them or come up short. They could not pass the increase on to consumers because of the retail price cap.

By January 2001, rolling blackouts were sweeping across the state, irritating home consumers and bringing businesses to a very expensive halt. At that point, California suspended deregulation and the Power Exchange, and the California Department of Water Resources replaced the three utilities as a power purchaser. In February, California governor Gray Davis began signing some $44 billion worth of long-term contracts with several energy suppliers, in which the state agreed to buy electricity from them at an inflated rate for terms that sometimes extended 20 years or more. In June, FERC finally stepped in (as the state had been begging it to do for months) and ordered a temporary price cap for wholesale electricity in all the western states.

Fortunately for California, a combination of cooler weather, conservation efforts, and a slowed state economy triggered in part by the electrical problems made demand drop during the summer. Wholesale electricity prices returned to pre-crisis levels. The state's pain was by no means over, however. The utilities were still locked into the expensive contracts the state had negotiated at the height of the crisis, and one utility, Northern California's Pacific Gas and Electric (PG&E), had to declare itself bankrupt in April 2001. PG&E was restructured and reemerged from bankruptcy three years later, and the state was eventually able to renegotiate some contracts and save itself about $7 billion, but California ratepayers were still left with an enormous bill—more than $1,000 per person, according to The Utility Reform Network, a California consumer advocacy group.

The "blame game" that had started during the crisis still continues. Some of the most substantial charges were laid at the door of Enron and other energy marketers and out-of-state power companies, whom Governor Davis called "pirates."[18] Evidence released as a result of lawsuits and FERC investigations in 2003 and 2004 bore out the term, revealing that the companies had deliberately taken power plants out of service and sent out false reports of congestion or overloading in transmission lines to drive up prices. Audiotapes made public in June 2004 showed Enron traders laughing about the money they were stealing from "Grandma Millie" in California.[19]

Several Enron executives, including Kenneth Lay, the company's former chairman, have been indicted or convicted on fraud charges, in some cases based partly on their activities during the California crisis. FERC also revoked Enron's marketing authority in June 2003, but this action meant little because the company's fraudulent actions had already forced it into bankruptcy in December 2001. In March 2003, FERC ordered Texas-based Reliant Resources and several other power companies to refund $3.3 billion to California for their part in the deceptions, but this was only a third of the amount the state had requested. Securities firm Morgan Stanley and El Paso Corporation, the country's largest gas pipeline company, have paid settlements to California as well, but they have admitted no wrongdoing.

Much as these companies' market manipulations may have added to California's problems, few commentators think they can bear the blame alone. Energy suppliers and traders and some analysts maintain that the state created most of its own problems with its badly designed deregulation policy. A special target for criticism was the combination of requiring utilities to buy on the volatile, expensive spot market and not allowing them to pass their costs on to consumers, which would have pressured the latter to decrease their demand.

Supporters of energy deregulation, including representatives of the Bush administration, have also claimed that, cheating aside, California did suffer a genuine shortage of electricity in 2000 and 2001. Much of this, they say, was due to environmentalists who insisted on stringent air quality regulations and middle-class citizens who blocked construction of new power plants in their neighborhoods (the so-called NIMBY, or "not in my back yard," syndrome). Other commentators, however, maintain that the supply would have been adequate if it had been purchased at reasonable rates and properly distributed.

There were plenty of other targets for criticism as well. Davis's response to the crisis, which his opponents pictured as slow and inadequate, played a part in forcing his recall from office and replacement by Arnold Schwarzenegger in late 2003. Other analysts have claimed that FERC's actions were equally tardy. Many observers say that all these parties should share in the blame.

The chief question left in the wake of the California crisis was whether it was likely to prove typical of state electricity deregulation. Those who think it was unique point to the greater success of Pennsylvania and Texas, whose deregulation policies have avoided some of the California legislation's pitfalls. In Texas, for instance, consumer prices are pegged to wholesale market prices, and generators and suppliers can make long-term contracts. These states also have more out-of-state electricity sources to draw upon than California did. Other observers, however, fear that the state's troubles will be typical or, at least, that investors will be frightened away by the thought that similar events might occur in other states.

Other countries, including Great Britain, Australia, and Argentina, also privatized and deregulated their formerly nationalized electricity industries during the 1990s. Their deregulation programs often shared features with those in the United States, including unbundling, reduction of regulation, creation of independent system operators and nationwide grids, and opening up of electricity generation and, to a lesser extent, marketing to competition. In 1996 the European Union issued a directive requiring open access to electric transmission, establishment of a transmission system operator, and financial unbundling of generation, transmission, and distribution. Member states were to have full retail competition in place by 2007. By July 2004, two-thirds of European electricity customers had at least a theoretical right to choose their energy suppliers.

Europe's experiments with deregulation have generally resulted in lower consumer prices for electricity. This has not necessarily been the case in the United States, however. A study that the National Center for Appropriate Technology conducted in 2002 concluded that restructuring generally had not produced the lower prices or increased consumer choice that regulators and legislators had hoped for. Indeed, in states that allowed electricity rates for small consumers to fluctuate with the market, consumers were found to be worse off than they were under traditional rate-setting policies or under deregulation with rate caps or freezes. Steven N. Isser wrote in the *Review of Policy Research* in 2003 that by 2000, many states were coming to feel that "deregulation had been oversold to consumers."[20] Supporters, however, say that true open competition in energy markets has not yet occurred in any state, so its effects cannot really be evaluated.

The Aging Transmission Grid

In summer 2003, a disaster more acute than the California crisis highlighted another electricity problem that, if not caused by deregulation, has certainly been worsened by it: the lack of reliability in the country's aging transmission grid, the network of wires that carries high-voltage electricity from generators to distributors. Most of the wires in the grid were put up in the 1950s and 1960s, when utilities generated their own electricity and transmitted it over relatively short distances. Conditions have changed in many ways since then, but the transmission system and its technology have not kept up with them. Spencer Abraham has stated that the U.S. grid is no better than that of a developing country.

The increased long-distance, often interstate, traffic and complex balancing of supply and demand from many different sources that occur in today's deregulated electricity environment frequently create potentially dangerous overloads. Reliability of the nation's electricity grid is increasingly important because more and more industry and daily activities depend on computers, which must have an uninterruptible, high-quality power supply. Even brief outages or changes in voltage or frequency can result in significant losses of information, productivity, and money for many businesses. FERC recommended in the late 1990s that each region of the country establish an independent organization to handle transmission management, either a for-profit transmission company, or transco, or a nonprofit company called an Independent System Operator (ISO).

When generating equipment shuts off, electricity from elsewhere in the grid surges into the "empty" area, potentially overloading and damaging lines and equipment. Conversely, when a transmission line breaks down, power washes back into the interconnecting lines, with the same result. As *New York Times* writers Matthew L. Wald, Richard Pérez-Peña, and Neela Banerjee put it, "The power grid is like a game of tug of war, which works as long as neither side—the generating stations and the load centers—wins. If one side falters, and the rope moves too far, everyone on the other side will fall down."[21] Relays and other equipment are supposed to shut off parts of the grid when abnormal flow is detected, in order to keep trouble from spreading, but they—and the humans who back them up—do not always succeed.

If both automatic and human grid monitoring fail, local problems can mushroom into regionwide ones in seconds. That happened on August 14, 2003, when what apparently began as a shutdown of generators in northeastern Ohio spread almost instantly throughout northeastern North America, leaving some 50 million people in New York, the Midwest, and Ontario, Canada, without power. Although fortunately this blackout, unlike an earlier one in New York City in 1977, was not marked by riots or looting, it was considered to be the worst in the nation's history.

As with the California energy crisis, many factors were blamed for the eastern blackout. An investigative panel convened by the governments of the United States and Canada stated in April 2004 that they had traced much of the problem to FirstEnergy Corporation, one of the largest utilities in the United States. The Ohio generators whose failure started the cascading disaster were in FirstEnergy's territory, and the panel concluded that the company should have responded to their shutdown by temporarily cutting off power to the Cleveland-Akron area in order to "shed load." Its failure to do so allowed the imbalance to spread, the panel said. The fact that management of the Ohio grid was divided between the Midwest Independent Transmission System Operator and PJM Interconnection also contributed to the blackout, because it meant that neither organization had full control of or complete information about the regional network. Neither, therefore, realized the extent of the problem or took sufficient steps to correct it.

No matter which groups are faulted for the events of August 14, most analysts agree that the overall condition of the transmission system made such an event—and quite possibly more like it—all but inevitable. Many blame the lack of maintenance and upgrading on deregulation, which has made utilities unwilling to invest in networks they no longer exclusively own or profit from. The electricity industry, in turn, complains that conflicting regulations and jurisdictions and the NIMBY syndrome make obtaining permits for new lines almost impossible.

GLOBAL ENERGY ISSUES

Increasingly, energy is a global issue. Energy and its sources are bought and sold in a world market, and a country's decisions about energy production and consumption can have major effects on far distant economies and environments. This is particularly true of oil and its often-associated hydrocarbon cousin, natural gas. The world's great powers recognize that access to these resources is essential to their economic well-being, and they shape their foreign policy accordingly.

OIL-PRODUCING COUNTRIES

Whether big or small, areas with large oil and gas reserves have a disproportionate impact on global politics. Unfortunately, most of these regions have been marked by civil and intercountry strife for centuries, if not millennia. (The only major exceptions are the United States, Canada, China, Norway, and Britain, the latter two of which co-own the considerable oil deposits in the North Sea.) Competition for oil has often made these disputes worse, particularly when several countries share the same resource.

Chief of these war-torn "hot spots," as everyone knows, is the Middle East, or Persian Gulf area. It holds a little more than half the world's proven oil reserves, according to an *Oil and Gas Journal* estimate published at the beginning of 2003. Furthermore, Mideastern oil is close to the earth's surface, which means it can be extracted much less expensively than oil from most other areas. Thus, the Middle East is, and is virtually guaranteed to remain, by far the world's most important source of petroleum. According to the Energy Information Administration, about a fifth of the oil that the United States imported in 2003 was from this region.

Saudi Arabia, which occupies most of the Arabian Peninsula, both produces and exports more oil than any other single country. It produced about 11 percent of the world's total in 2002. Because Saudi Arabia (unlike most other countries) has production capacity that can be easily started or stopped, it is the world oil market's "swing producer," or market regulator, determining spot market prices almost singlehandedly by changing the amount of oil it exports.

Realizing Saudi Arabia's importance, the United States has maintained close ties with the country's ruling family since World War II. That alliance has helped to maintain the Saudi government, not only against possible attack from outside, but against the threat of rebellion by Muslim fundamentalists such as Osama bin Laden's al-Qaeda organization. Saudi Arabia has avoided civil war so far, but terrorist attacks in the country during May and June 2004, including the beheading of an American contractor, caused spikes in oil prices because of fears that the government was no longer able to protect its trading partners. "The entire world economy is built on a bet of how long the House of Saud can continue," Philip E. Clapp, president of the National Environmental Trust, claimed in early 2003.[22]

Most of the Middle East's other major oil-exporting countries have at least equally rocky recent histories. The net oil exports of Iran, the United Arab Emirates, Iraq, and Kuwait ranked fifth through eighth, respectively, in the world in 2000. An anti-American fundamentalist government took control of Iran in 1979, and the United States has forbidden U.S. companies and their foreign subsidiaries to do business there since 1995. Iraq invaded Kuwait in August 1990, and in return the United States invaded Iraq the following January. It did so again in March 2003 and overthrew the country's dictator, Saddam Hussein, later in the year. Only the Emirates have been relatively peaceful.

Russia is the world's second-largest producer and exporter of oil. It also has the world's largest natural gas reserves, second-largest coal reserves, and eighth-largest oil reserves. Most of the country's economic growth in the first years of the 21st century has been fueled by increases in its oil exports. The bulk of the oil comes from western Siberia, but several other parts of the country also are, or are likely to become, important sources, including the Caucasus, Sakhalin Island (just north of Japan), and the northern part of the Caspian Sea, a landlocked body of water north of the Mediterranean. Russia presently ships oil primarily to northern Europe but is considering pipelines that would send this resource to Asia as well. In spite of its immense reserves, Russia exports little natural gas.

The Soviet Union was a major oil producer during most of the 1980s, but that sector was neglected after the Soviet government collapsed in 1991. Russian premier Boris Yeltsin privatized the country's oil industry in the early 1990s, and aggressive entrepreneurs such as Mikhael Khodorkovsky bought Russian oil companies and have managed them ever since. These "oligarchs," as they have often been called, remodeled the companies along Western, competitive lines in the late 1990s and were probably responsible for the Russian oil industry's renewed success, but they have not been popular with the country's people or government. Some observers think that Khodorkovsky's arrest on fraud charges in October 2003 signals the government's desire to reassert control of the oil companies, perhaps to the extent of renationalization.

The region of Russia that presents the greatest global concern, both because of its apparently rich oil and gas deposits and because of its potential for international disputes that could rival those in the Middle East, is the part that borders on the Caspian Sea. The EIA claims that by 2010 this area may produce at least as much oil as Venezuela, the world's eighth-largest producer in 2002, although some other analysts say that its potential has been exaggerated. The sea is bordered by Russia on the northwest, Iran on the south, and three independent countries that were formerly parts of the Soviet Union—Kazakhstan, Turkmenistan, and Azerbaijan—on the remaining sides. Two other former Soviet states, Uzbekistan and Georgia, also play roles in Caspian politics because parts of proposed pipeline routes from the area's oil fields cross their territory.

After several large oil discoveries around the Caspian were made in the 1990s, Western Europe and the United States began looking to the area as a possible source of oil to replace the troubled Middle East. They have increased their military as well as diplomatic and economic presence there, but Russia wants to keep control of the area if it can. This competition has led analysts such as Michael Klare of Hampshire College in Massachusetts to call the Caspian the site of a new "Great Game," reminiscent of the 19th-century struggle between Britain and czarist Russia over the same territory.

The New Great Game is by no means the Caspian's only power struggle. The small countries bordering the sea—all very poor and with governments weak at best and dictatorial at worst—contain numerous religious and ethnic groups that have been fighting throughout history. "The political and racial divisions of this kaleidoscope make even the Middle East look almost harmonious," Sarah Searight wrote in an article about the region in *History Today*.[23] (Because Iran controls the area's southern border, conflicts in the Caspian in fact potentially overlap those in the Middle East.) In addition to these internal struggles, the countries disagree about the exact location of their borders, including those under the sea itself, where most of the oil deposits lie. Equally thorny political problems, as well as physically difficult terrain, affect siting of pipelines to transport oil from the landlocked Caspian to seaports from which it can be shipped.

Latin America, too, has its share of oil. Venezuela and Mexico are its most important oil-producing countries, with Mexico ranking sixth and Venezuela eighth in world oil production in 2002, according to the EIA. Among oil exporters, however, Venezuela was fourth and Mexico only 10th in 2000, probably because Venezuela exports a larger share of its oil than Mexico does. In 2003, Mexico was the third-largest and Venezuela the fourth-largest source of oil imported into the United States, exceeded only by Canada and Saudi Arabia. Both countries have been producing oil since the early 20th century and have nationalized their oil industries. Venezuela is the only non-Arab nation that has belonged to the Organization of Petroleum-Exporting Countries (OPEC) since its founding. It has seen many internal disputes over oil, including a month-long strike in December 2002 that temporarily shut off the country's oil exports, greatly damaged its economy, and nearly toppled the presidency of Hugo Chávez.

Sub-Saharan Africa, although relatively new to the ranks of oil producers, is another area likely to become an increasingly important source of petroleum supplies—and a locus of potential major conflicts. Nigeria, the continent's largest producer, ranked ninth among world oil exporters in 2000. Other important oil-producing areas are Angola, Gabon, the Congo, Equatorial Guinea, Sudan, and, most recently, the central African nation of Chad. Ian Gary of Catholic Relief Services wrote in mid-2003 that "West Africa's [oil industry] growth potential is considered to be greater than that of Russia, the Caspian or South America."[24] Most of these countries, like many other oil-rich developing nations, are marred by poverty, religious and ethnic disputes, and in some cases outright civil war.

In addition to oil-producing countries themselves, sources of potential conflict over oil include so-called choke points through which large shipments of petroleum must pass. In 1999, the U.S. Department of Energy listed six such choke points, which collectively affected more than 40 percent of the oil consumed in the world: the Strait of Hormuz, which connects the Persian Gulf to the Indian Ocean; the Strait of Malacca, between

Malaysia and the Indonesian island of Sumatra, which connects the Indian Ocean to the South China Sea; Bab el Mandeb, at the mouth of the Red Sea; the Suez Canal and the Sumed Pipeline, which connect the Red Sea to the Mediterranean; the Bosporus/Turkish Straits, which connect the Mediterranean to the Black Sea; and the Panama Canal, which connects the Atlantic to the Pacific Ocean. A terrorist attack or a war in any of these areas could have devastating effects on world oil prices and supplies.

National and Multinational Organizations

Many oil-producing countries have placed their petroleum assets in the hands of national companies, often run by the countries' governments. Some of these companies, such as Saudi Arabia's Aramco, are very large. Wayne Ellwood wrote in the *New Internationalist* in June 2001 that the top 10 national oil companies controlled 70 percent of the world's reserves at that time. Some national oil companies do not allow foreign investment, but others, especially in poorer countries, encourage it. In those cases, in return for supplying the expensive equipment and high-tech expertise that modern drilling projects need, multinational oil companies, mostly privately owned and based in the United States or Europe, become partial owners or controllers of the countries' energy assets.

In 1970 the then-largest oil companies were known as the "Seven Sisters": Royal Dutch/Shell, Exxon, Gulf, Texaco, BP, Mobil, and Standard Oil of California (Chevron). Royal Dutch/Shell and BP (formerly British Petroleum) began as European companies, while the others were based in the United States and descended, directly or indirectly, from John D. Rockefeller's late 19th-century behemoth, Standard Oil. Like their ancestor, these companies were vertically integrated. At that time, the Seven Sisters had what Wayne Ellwood called "a stranglehold" over the world's oil trade, including refined products as well as crude oil.[25]

The "sisters" also had a rival, however, which was soon to show its power: the Organization of Petroleum-Exporting Countries, or OPEC. Formed in 1960 to "coordinate and unify petroleum policies among Member Countries, in order to secure fair and stable prices for petroleum products," OPEC originally consisted of Saudi Arabia, Iran, Iraq, Kuwait, and Venezuela.[26] Then and now, OPEC members attempt to control the price of oil by agreeing on how much oil each member will export during a given period.

OPEC had little influence during the 1960s, partly because of a worldwide oil glut. In the early 1970s, however, a drop in U.S. oil production, new members joining the organization, and closer cooperation among member countries greatly increased OPEC's ability to affect oil prices.

It—or, rather, a sub-organization within it called the Organization of Arab Petroleum Exporting Countries, or OAPEC—used that influence most famously in October 1973, when it agreed not to sell oil to the United States, Portugal, and the Netherlands because these countries had supported Israel in the brief Arab-Israeli war that began on October 6. Although the Arab oil embargo lasted only until March 1974, this employment of what came thereafter to be called the "oil weapon" had a profound impact on the U.S. economy and psyche, as people waited in line for hours to buy gasoline. Indeed, the effects were worldwide.

Another OPEC member employed the oil weapon again in 1979, when Islamic fundamentalists ousted Mohammad Reza Pahlavi, the pro-Western shah of Iran. In November, after the new government seized and held several Americans in the country as hostages, President Jimmy Carter banned the importation of Iranian oil, and Iran, in turn, levied another oil embargo on the United States. Fueled by memories of the gas lines of 1973–74, panic buying sent oil and gasoline prices soaring again. World prices peaked in 1981, then went into a steady slide that, by 1985, had had almost as painful an effect on the world economy as the sudden rises had produced. OPEC tried to respond by reducing production, but its member countries could not agree on what cuts should be made or by whom.

As of 2004, OPEC has 11 member countries: Algeria, Indonesia, Iran, Iraq, Kuwait, Libya, Nigeria, Qatar, Saudi Arabia, the United Arab Emirates, and Venezuela. Together they account for about 40 percent of global oil production. Although the "oil shocks" of the 1970s led Americans to fear and revile OPEC, many commentators say that this group, although still an important factor in the oil market, no longer has the power to do major economic harm even if it wanted to. For one thing, the United States and other developed countries have made themselves less vulnerable to OPEC actions by diversifying their sources of imported oil, building strategic oil reserves, and learning to use petroleum more efficiently.

OPEC also has some rivals. Just as it was formed in response to what its members saw as excessive control by the multinational oil companies, the 16 countries most affected by the first Arab oil embargo, including the United States, met in 1974 to create the International Energy Agency (IEA) as a counterbalance to OPEC. IEA members agreed to each hold oil stocks representing 60 days' worth of their net imports, later raised to 90 days, and worked out a system to allocate those stocks if oil supply fell below demand by more than 7 percent. As with the U.S.'s Strategic Petroleum Reserve, different countries have different opinions about when the reserves should be drawn down. The European Commission, for one, wants the IEA to order the release of stocks if world oil prices become extremely high, but the IEA's policy is to call on them only in case of supply disruption. So far, the agency has asked for release of member stocks only once, briefly, at the start of the 1991 Gulf War with Iraq.

The multinational oil "sisters" are also regaining strength, although mergers have made their family smaller. Exxon merged with Mobil in 1999, for example, and Chevron with Texaco in 2001. Wayne Ellwood wrote in 2001 that, after decades of being partly overshadowed by OPEC and weakened by competition from nationalized and small independent companies, the multinational giants were "clawing their way back to power" and were "amongst the most powerful, influential corporations in the world."[27]

The most important reason why OPEC's "oil weapon" is blunted today, however, is that the price of oil can no longer be controlled by any one company, country (with the possible exception of Saudi Arabia), or group of countries. Instead, it is now set—and rapidly changed—in world commodity spot and futures markets such as the New York Mercantile Exchange and the International Petroleum Exchange. National Defense University professor of economics Donald Losman wrote in early 2002, "The entire world 'drinks' from what is essentially one global [oil] pool; any refusal to sell to the U.S. while supplying other countries simply releases some other seller's oil to be sold to America."[28]

OIL, WAR, AND NATIONAL SECURITY

Oil has literally fueled wars, and sometimes helped to cause them, throughout the 20th century. Oil became a significant factor during World War I because in 1912 a young Winston Churchill, as First Lord of the Admiralty, changed the fuel of the British navy from coal to oil. Because Britain had very little of its own oil, it obtained the needed supplies by acquiring a majority stake in the Anglo-Persian Oil Company, which had discovered oil in Persia (now Iran) a few years before—the first oil-related activity in the Middle East.

During the war, Britain's oil-powered ships proved faster and more dependable than Germany's coal-powered ones. Planes, tanks (introduced in 1916), and numerous other land-based vehicles also used fuel made from petroleum. Lord George Nathaniel Curzon, later Britain's foreign secretary, said afterward that Britain and its allies had "floated to victory upon a wave of oil."[29] With this lesson in mind, Britain expanded its area of control over oil-rich land in the Persian Gulf in the years following the war—and recognized protection of this land as an important part of its own national security. Other countries, including Germany, France, and Japan, also set up government-owned oil companies and staked out foreign sources of supply. The United States established ties with Iran and Saudi Arabia.

Oil helped to shape events in World War II as well. Germany attempted to invade the Soviet Union partly to obtain oil from the Caspian area. Japan's aggression in Southeast Asia was also spurred largely by its desire to take over oil-containing lands in what were then the Dutch East Indies. The failure of both efforts helped to lead to the countries' defeat, as did the Allies' possession of ample oil stocks. Echoing Lord Curzon's words about the earlier worldwide conflict, Russian premier Joseph Stalin made this toast near the end of the war: "This is a war of engines and octanes. I drink to the American auto industry and oil industry."[30]

As Britain slowly removed itself from the Middle East in the 1960s, the United States, working through rulers such as the shah of Iran and Abd al-Aziz ibn Saud, the founder of Saudi Arabia, took its place as a major foreign influence in the area. At first the main U.S. policy aim there was the cold war goal of keeping the region out of Soviet hands. However, after the United States became a net importer of oil in the early 1970s and the decade's Arab oil embargoes sent shock waves through the economy, the U.S. government came to see protection of its own and its allies' access to Middle Eastern oil supplies as essential. In 1980, after the second "oil shock," President Jimmy Carter told Congress, "An attempt by any outside force to gain control of the Persian Gulf region will be regarded as an assault on the vital interests of the United States of America . . . [and] will be repelled by any means necessary, including military force."[31]

This policy, which became known as the Carter Doctrine, spawned a steady buildup of U.S. military forces in the Middle East during the 1980s. It was also used at least in part to justify the U.S.-led war against Iraq that began in January 1991, in response to Iraq's invasion of its small but oil-rich neighbor Kuwait, in August 1990. Daniel Yergin, author of *The Prize*, a history of the oil industry, claims that "[Iraqi leader] Saddam's objective [in invading Kuwait] was, as it had been when he invaded Iran a decade earlier, to control, directly or indirectly, a large part of Persian Gulf supplies and bend them to his political and military purposes."[32] In turn, Michael Economides and Ronald Oligney wrote, "The message [of U.S. action in the Gulf War] was unmistakable: Disturbing or threatening to disturb petroleum supplies and trading would not be tolerated."[33] Or, as the always-quotable Amory Lovins and Hunter Lovins put it, "There was more at stake in the Gulf War than just oil; but we'd hardly have sent half a million troops there if Kuwait just grew broccoli."[34]

Defense of access to oil, especially in the Middle East, has remained an apparent cornerstone of U.S. foreign policy ever since. It was thus not surprising that antiwar protesters claimed that oil was the chief, if not the only, reason for the U.S. invasion of Iraq that began in March 2003. Some political analysts agreed with them. Bush administration spokespeople and oil company representatives ridiculed this allegation, however. Secretary of Defense Donald Rumsfeld, for instance, insisted in a radio interview on November 15, 2002, that the possibility of invading Iraq had "nothing to do with oil."[35] Similarly, J. Robinson West, chairman of PFC Energy in Washington, D.C., wrote in mid-2003, "Oil companies were not the drivers behind the war. . . . The last thing the oil companies want is instability in financial markets."[36] Even if seizing Iraqi oil had been the war's aim, oil experts say, it would have been a foolish one, because controlling Iraq would not have a major effect on the world market in oil.

The "war on terrorism" has also been tied to oil. Michael Klare has claimed, "The merger of the war against terrorism and struggle for oil . . . provide[s] the White House with a convenient rationale for extending U.S. military involvement into areas that are of concern to Washington primarily because of their role in supplying energy to the United States."[37] Klare mentions military actions in the Caspian area and Colombia as well as in Iraq. Lita Epstein and the other authors of *The Complete Idiot's Guide to the Politics of Oil* also point out that Afghanistan, invaded after the terrorist attacks of September 11, is the potential site of one of the proposed Caspian pipelines.

To be sure, U.S. foreign policy during this same period has also tried to reduce the country's dependence on Middle Eastern oil by diversifying the sources from which the United States imports oil, especially increasing the proportion of oil imported from other countries in the Americas. As a result, contrary to public perception, most of the oil that the United States imports—an estimated 58 percent of its total consumption in 2004—does not come from the Middle East. The country's largest single supplier in 2004 was Canada. Mexico was second, Saudi Arabia third, and Venezuela fourth.

The United States is not the only country that has been accused of going to war for oil. Michael Klare, among others, claims that interest in a potential pipeline route from the Caspian to the Black Sea lies behind Russia's determined attempts to keep control of Chechnya and Dagestan, including its 1999 invasion of Chechnya. China is pressing equally firmly on its rebellious province of Xinjiang, a potential source of oil for that energy-hungry country. Ed Ayres, writing in *World Watch* in mid-2003, maintained that oil contributed to civil wars or suppression of rebellions in Angola, Indonesia, Sudan, Chad, Burma, Venezuela, Russia (western Siberia), Colombia, and China (Tibet). "Oil lubricates wars that got off to creaky starts for other reasons," Ayres wrote. "It pays for escalation and perpetuation, and after a while the oil revenue can become the main object even if it wasn't originally."[38]

Finally, oil can encourage terrorism in several ways. Money from oil-producing countries' sales can be used to fund terrorist groups, and resentment of actions taken by oil companies or developed countries' governments can help to draw in new recruits. Oil tankers, pipelines, storage facilities, and refineries also make tempting targets for attacks. "The deadly trinity of oil, war and dictatorship presents the greatest challenge to humanity at the start of the new millennium," writes John Bacher, past president of the Preservation of Agricultural Lands Society in Canada, in his book *Petrotyranny*.[39]

Social Problems and Human Rights Abuses

"Striking it rich" with a large oil find might seem a dream come true for a previously poverty-stricken developing country, but numerous economists and social scientists have concluded, based on the experiences of nations such as Nigeria and Venezuela, that it is more likely to be a nightmare. They say that mature democracies such as Britain and Norway can handle sudden oil wealth because they have mechanisms to ensure fairly responsible government spending and to obtain input from citizens. For countries in which democracy has a recent and tenuous hold or no hold at all, however, an influx of money from oil sales seems almost guaranteed to lead to corruption, abuse, and a growing financial gap between the country's elite and the rest of its people.

The reason why oil can be a curse for such countries, say analysts such as Moises Naim, editor of *Foreign Policy* magazine, is that a government's wealth normally comes chiefly from taxes. Even in a country ruled by a dictator, taxation requires some degree of consent. If citizens object strongly enough to the amount of taxes or the uses to which tax revenue is put, they may rebel and even overthrow the government. Money from the sale of oil or other natural resources, however, comes from outside the country or from a few large companies within it rather than from the bulk of the country's people. The government therefore does not feel obliged to share that wealth with them. Instead, it concentrates on pleasing the sources of the wealth and benefitting itself and its friends. It grows ever more dictatorial and corrupt, and its citizens become poorer and more abused. Oil industry in such countries, furthermore, tends to overwhelm and cripple other economic sectors, creates few jobs for local people, and is subject to boom-bust cycles as world oil prices fluctuate.

Although multinational oil companies do not create or directly control social and political conditions in the countries where they do business, they have frequently been accused of making these conditions worse. For instance, Wayne Ellwood claimed in *The New Internationalist* in June 2001, "the oil giants move into environmentally fragile and isolated frontier areas . . . with scant consideration for local cultures or the human-rights records of governing regimes."[40] In the April 28, 2003, edition of *Forbes*, Daniel

Fisher wrote, "There are plenty of Saddamlets around the world with whom . . . every . . . Western oil company openly does business. They have little choice."[41]

Some of these accusations have ended up in the courts. For example, after a three-year Justice Department investigation—the largest American investigation into alleged bribery in other countries—James Giffen, CEO of a New York merchant bank called Mercator Corporation, was indicted in a New York court in April 2003 for allegedly paying $78 million in bribes to the president of Kazakhstan, Nursultan Nazarbayev, and officials in his administration in the 1990s on behalf of ExxonMobil, ConocoPhillips, Amoco (now part of BP), and possibly ChevronTexaco. None of these companies was directly charged with wrongdoing, but a former Mobil executive, Bryan Williams, was also indicted. Both men declared their innocence, and as of mid-2004, the case was still pending.

In the Giffen case and some similar ones, oil companies or their representatives have been charged with violating the Foreign Corrupt Practices Act, a 1977 law that forbids U.S. citizens or corporations from paying bribes to foreign officials. The United States may be the only country to have such a law, but in 1997 the members of the Organization for Economic Cooperation and Development, which includes most of the world's industrialized (and many developing) countries, signed the Convention on Combating Bribery of Foreign Public Officials in International Business Transactions, which also makes bribing a foreign official a criminal offense.

Human rights groups have turned to a far older law, passed in 1789, to attack oil companies that they say abetted rights abuses in developing countries. This law, the Alien Tort Claims Act (ATCA) or Alien Tort Statute (ATS), was originally intended to allow a small number of cases involving certain violations of international law, probably mainly piracy and attacks on ambassadors, to be heard in U.S. (federal) courts. In 1980, however, an appeals court ruling in a case called *Filartiga v. Peña-Irala* permitted ATCA to be applied to cases of torture and genocide, which were held to be international crimes. By the end of 2003, more than two dozen ATCA suits had been filed, including several against oil companies.

Kenny Bruno, campaigns coordinator for EarthRights International and co-counsel in some of these suits, wrote in *Multinational Monitor* in March 2003, "ATCA gives hope to victims that any U.S. corporation complicit in grave human rights abuses anywhere in the world can be held accountable in U.S. courts."[42] Both business associations and some policy analysts, however, disagree with this use of the statute and would like to see it amended or repealed. The International Chamber of Commerce says that human rights suits citing ATCA are "an unacceptable extraterritorial extension of U.S. jurisdiction" and warns that they could reduce both U.S.

business investments abroad and foreign companies' investments in the United States.[43] Daniel Griswold of the libertarian Cato Institute says that ATCA suits are "bad law, bad economics and bad foreign policy."[44] A June 2004 Supreme Court decision in the case of *Sosa v. Alvarez-Machain* made successful future use of ATCA in human rights suits against corporations unlikely, though not impossible.

ATCA was also invoked in a suit against ChevronTexaco, first filed in 1999, following a 1998 incident in which about 100 protesters seized one of the company's offshore oil rigs and held some 200 of its workers hostage for three days. Nigerian government forces put down the protest, killing two Nigerians in the process. The suit, filed on behalf of members of the families of the two slain men, alleged that the men were merely protesting and that Chevron Nigeria officials ordered government security forces to shoot them, as well as providing helicopters for the forces. In late March 2004, a district judge ruled that the plaintiffs had presented enough evidence to allow ChevronTexaco itself, not just its Nigerian subsidiary, to be sued, meaning that the suit could continue in the United States.

The International Labor Rights Fund, EarthRights International, and other human rights groups called upon ATCA in federal suits against the U.S. oil company Unocal for abetting alleged abuses in Burma (Myanmar) during the late 1990s. Burmese villagers claim that the country's military government routinely used forced labor ("We were slaves," one said bluntly) on construction of a natural-gas pipeline in which a Unocal subsidiary had a minority interest.[45] The rights groups maintain that Unocal was, or should have been, aware of this and yet made no effort to stop it, a charge the company denies. In a related state suit, California Superior Court judge Victoria Chaney ruled in June 2002 that Unocal was not directly liable for the abuses but might be held vicariously, or indirectly, liable. She further ruled in early 2004 that the groups needed to file suit against Unocal's Burma subsidiary, not the parent company, because the subsidiary did not qualify as an "alter ego" of the parent. In late 2004, both suits were settled out of court on undisclosed terms.

The World Bank and some other international organizations that provide financing for projects in developing countries have conditioned their loans upon companies filing social and environmental impact reports for proposed oil and gas projects and taking steps to mitigate the harm that such projects may wreak. In some cases they have appointed or required the hiring of independent groups to prepare the reports or oversee the mitigation programs. The World Bank Group did so, for instance, for a $3.5 billion loan to the governments of the neighboring Central African countries of Chad and Cameroon and a consortium of oil companies for building an oil pipeline from southern Chad to the Cameroon coast. However, William

Jobin, director of Blue Nile Associates and a member of the panel, claimed in June 2003 that the group's advice "was largely ignored by the project proponents, and had little success in minimizing the most serious impacts or in improving the social equity of the project," a situation he says the World Bank did nothing to rectify.[46]

Oil companies and international lending agencies say they are trying to improve their human rights record, and several have sponsored social programs in countries where they do business. ChevronTexaco has cosponsored a number of education, training, and small-business-development programs in Angola, for example. To provide increased "transparency" for their actions, several oil companies, including ChevronTexaco and Royal Dutch/Shell, also publish annual corporate responsibility reports, which include self-criticism.

ENVIRONMENTAL EFFECTS

Lawsuits have been used to protest alleged environmental as well as social damage from oil and gas projects. A group of indigenous people in Ecuador, for example, with the help of outside environmental and human rights groups such as Amazon Watch, filed suits in the United States to recover damages they say are due to them because projects comanaged by Texas Petroleum, a ChevronTexaco subsidiary, and PetroEcuador, Ecuador's national oil company, dumped 18.5 billion gallons of toxic waste water into local waterways in the 1970s and 1980s. Lawyers for ChevronTexaco have said that they spent $40 million in the late 1990s to clean up the area and that any remaining contamination is due to PetroEcuador's continuing activities. A federal appeals court ruled in 2002 that the case should be tried in Ecuador rather than the United States. It opened there in 2003 and, as of early 2005, was still proceeding.

Amazon Watch, Friends of the Earth, and other environmental groups have also tried to stop Peru's Camisea Project, which involves a pipeline to carry natural gas from the Amazon rain forest to Lima, the country's capital, for domestic use and to the coast for conversion to liquefied natural gas and other exportable forms. The Lima portion of the pipeline began carrying gas in August 2004, and the gas liquefaction plant was to begin construction at about the same time and start exporting around 2008.

The Camisea Project should bring badly needed revenue to Peru, but environmentalists say it could also spell disaster for pristine environments and native groups. They point out that the Camisea gas field is near two prime biodiversity spots the Apurimac Reserve and Manu National Park, and the export terminal is near the Paracas marine reserve, so spills or other accidents connected with the project could seriously damage these valuable areas. The companies working on the project have said they are taking steps to protect the environment, such as bringing supplies to the drilling area by helicopter rather than building roads through the forest. Energy and Mines Minister Jaime Quijandria claimed in June 2004 that the Peruvian government had complied with all requirements to prevent environmental damage near the gas liquefaction plant.

Disturbing as environmental damage to particular areas may be, it is dwarfed by the possible damage to the entire planet posed by global warming. Because this problem affects the whole world, most analysts feel that solving it will require coordinated international activity. The chief attempt to mandate such activity so far has been the Kyoto Protocol, an agreement drafted in Kyoto, Japan, in December 1997 and signed by representatives of about 180 nations. This agreement, in turn, grew out of the 1992 Earth Summit at Rio de Janeiro, Brazil, where member countries committed themselves to the Framework Convention on Climate Change. The convention required "stabilization of greenhouse gas concentrations in the atmosphere at a level that would prevent dangerous anthropogenic [human-caused] interference with the climate system."[47]

The Kyoto Protocol calls for the world's 38 most industrialized countries to reduce their carbon-based gas emissions to an average of 5 percent below their 1990 levels by 2012. (Developing countries were not required to reduce their greenhouse gas emissions because industrialization was held to be so vital to their economies.) At the Kyoto meeting, for instance, the United States agreed to a 7 percent cut in emissions and the European Union to an 8 percent cut.

Details of implementing the protocol were worked out at Bonn, Germany, and Marrakesh, Morocco, in 2001. They included establishment of a cap-and-trade scheme something like the one the United States has used to reduce the sulfur dioxide emissions that contribute to acid rain. In this arrangement, nations that produce less than their targeted amount of emissions can sell "assigned amount units" to countries that produce more than their share. Some individual countries, such as Britain and Denmark, have set up their own cap-and-trade systems for greenhouse gases as well, and the European Union is planning to start a union-wide Emissions Trading Scheme, covering about 40 percent of greenhouse gas emissions from 20 countries, in 2005.

Some countries and international groups, particularly Britain and the European Union, have embraced the Kyoto Protocol enthusiastically. By contrast, although the United States ratified the Framework Convention in 1992, helped to draft the Kyoto document in 1997, and signed it in 1998, it has refused to ratify the protocol. The U.S. government claims that meeting the treaty's requirements would seriously damage the country's economy. It has also protested the protocol's lack of restrictions on developing countries, which are expected to produce greater increases in CO_2 emissions than developed countries in future years. President George W. Bush has said that he favors a voluntary program to reduce carbon dioxide emissions per unit of economic production as an alternative to the protocol.

The Kyoto Protocol becomes binding international law when 55 countries (the so-called Annex I parties), producing collectively at least 55 percent of world carbon dioxide emissions in 1990, ratify it. Thirty-two Annex I parties, accounting for 44.2 percent of the required emissions, had ratified the treaty by mid-2004, but China and Australia as well as the United States had refused ratification. The world therefore looked to Russia as the country whose agreement could bring the treaty into effect. In October 2004, after years of hesitation, both houses of the Russian Parliament approved ratification of the Kyoto Protocol, and President Vladimir Putin signed the ratification on November 4. Russia's approval meant that the Kyoto treaty went into effect on February 16, 2005.

THE FUTURE OF ENERGY

There seems to be little doubt that the world's demand for energy will go on rising in the coming decades. In a forecast released in April 2004, the Energy Information Administration predicted that world energy consumption would increase by 54 percent between 2001 and 2025, with most of that rise coming from developing countries as they grow. Developing Asia, especially China and India, alone is expected to account for 40 percent of the total increase and 70 percent of the increase projected for the developing world. Thanks to relatively slow population growth, increases in energy efficiency, and a move from manufacturing to service industries, energy use in developed countries is predicted to rise by only 1.2 percent per year. Because of these changes, the gap in energy consumption between the developing and the developed world is expected to narrow to the point of almost vanishing.

ENERGY SOURCES

The king of energy sources today is oil, but that will not always be true. In a sense, almost everyone agrees that the world is running out of oil; only a finite amount of that valuable substance was formed, and the global population is using it up at a profligate rate. There is no agreement at all, however, on when the last oil will be used up or, a date that many experts say is more important, when world oil production will peak. After this time, oil will become ever scarcer and more expensive, and—unless conservation or use of other energy sources substantially reduces demand for it—competition to obtain it will grow fiercer.

Pessimists say that world oil production will peak in the first decade of the 21st century, if indeed it has not already done so. Most base their opinions on the work of U.S. geophyicist M. King Hubbert, who concluded in the 1950s that oil production over time followed a bell curve. Hubbert became famous because he predicted in 1956 that oil production in the United States would peak around 1970—and it did. Applying an updated version of Hubbert's calculations to past world oil discovery and production and estimated reserves, Princeton University's Kenneth Deffeyes, a geologist who worked with Hubbert in the late 1950s and 1960s, has concluded that the world's peak oil production years are now occurring or perhaps are even past. (Peak production can be detected only after the fact, when production begins to decline.) Renowned British oil geologist Colin J. Campbell also thinks that the peak will come somewhere in the 21st century's first decade.

Other commentators—including the U.S. Geological Survey as well as the EIA—insist that production of oil and other fossil fuels will peak much later and that these fuels will remain dominant among energy sources for at least another 25 years. "People who are predicting an imminent peak are simply wrong," says Boston University economist Robert Kaufmann.[48] These optimists generally place the oil peak 50 to 100 or more years in the future. They point out that geologists have been warning about impending shortages since at least the 1920s but have always been wrong because improvements in technology enabled the discovery of new supplies or the economical extraction of old ones. These hopefuls think this will continue to happen.

Those who put their faith in technology point, for example, to recent discoveries of vast pools of oil beneath the floor of the deep sea, especially in the Gulf of Mexico and off the coasts of West Africa and Brazil. Thanks to new forms of drilling rig, oil can now be extracted from deposits as much as two miles below the ocean's surface and another 6.5 miles beneath the ocean floor. To be sure, these new technologies bring with them new risks: The ecological consequences of a deep-sea oil spill or well blowout are unknown but could be devastating.

Some types of previously known oil deposits may also be worth a second look. Oil is considered not worth extracting when the amount of money or energy needed to remove a given amount of it is greater than the amount of money or energy that amount will yield. Thus, some conventional oil fields have been classified as "depleted," even though they still contain substantial amounts of oil. New extraction techniques have reduced the expense of removing this residual oil, so some companies are returning to old fields for a new harvest.

New techniques are also making geologists reconsider other types of deposits whose oil was formerly thought impossibly difficult to extract. One of these is so-called tar sands or oil sands, of which the Canadian province of Alberta has a tremendous amount. In tar sands, grains of sand are surrounded by tiny envelopes of water that are in turn coated with tarry bitumen, a form of petroleum. A technique called steam-assisted gravity drainage, in which steam is injected into a horizontal well drilled above a second well, is beginning to make extraction from tar sands feasible. Heat from the steam melts the bitumen, and gravity drains it into the well below. Environmentalists point out, however, that this is a tremendously destructive process. The sand must be strip-mined in great quantities from the earth's surface (two tons of it are needed to yield a single barrel of oil), and it leaves behind many tons of used sand and huge ponds of polluted waste water.

Large deposits of a second previously rejected source, oil shale, exist in Wyoming, Utah, and Colorado. Oil shale contains organic matter called kerogen, which potentially can be converted to oil. The rock must be heated to high temperatures and treated in several other ways to extract useful petroleum from it, however. Skeptics say that past efforts to pull oil from oil shale have always been economic failures, and they think this will continue to be true. Removing the rock by strip-mining also would destroy beautiful Rocky Mountain landscapes, and processing it would use large amounts of scarce water. Finally, heated oil shale expands like popped popcorn, so the used rock cannot be disposed of by returning it to the holes from which it was removed.

One source of disagreement about how long world oil supplies will last centers on the calculation of so-called proven reserves, because all estimates of future production depend in part on this figure. Proven reserves are supposed to be quantities of unextracted oil or gas known to exist and believed to be removable with present technology under current economic conditions. Oil companies and oil-producing countries list their proven reserves yearly.

Analysts who say that the world will not run out of fossil fuels any time soon point to the fact that figures for proven reserves periodically jump upward as new technology and discoveries move more fuel into this category. Critics, however, say that "proven reserves" are often anything but. There is no agreed-on formula for determining them, so different analysts arrive at different results. Calculations may under- or overestimate the effects of technological improvements, for instance. Doubters of reserves figures' accuracy point out that companies and countries have a strong motive to exaggerate them because doing so pleases existing investors, attracts new ones, and increases influence in oil markets. Evidence that such exaggeration occurs was provided in January 2004, when multinational giant Royal

Dutch/Shell admitted that it had overstated its proven oil reserves by 20 percent.

Natural gas is another fossil fuel that may soon be in short supply—or not. The EIA predicts that demand for gas will grow faster than for any other primary fuel, 67 percent between 2001 and 2025, mostly because of increased use for electricity generation. The agency expects gas to pass coal as the most common fuel in electric power plants by 2010. Because of this increased demand, gas, like oil, could soon enter a phase of decreasing supplies and rising prices unless science saves the day by finding ways to extract it from new types of deposits. Some experts say that methane, the chief chemical in natural gas, exists in virtually inexhaustible quantities in icelike compounds called hydrates, found in permafrost and the deep sea. Techniques for releasing methane from hydrates are still in their infancy, however.

Even if gas supplies are sufficient, transporting them can be difficult. Many observers think that much of the gas that the United States uses in the future will come from Canada and Alaska's North Slope, which will require construction or expansion of extensive pipelines. Alternatively, gas can be turned into a liquid by lowering its temperature to –259°F, and this liquefied natural gas (LNG) can be transported in special tanker ships to ports constructed to receive them. Once there, the LNG is warmed to return it to a gaseous form, then shipped by pipeline to its final destination. Valuable as LNG might be, no one seems to want it anywhere near them. Environmentalists say that a tremendous explosion and fire could result if an LNG tanker or storage facility is ignited by either an accident or a terrorist attack. Over and over, worried residents have vetoed proposed sites for LNG tanker terminals in California and the Gulf Coast. Nonetheless, the Energy Information Administration expects LNG imports to increase substantially during the coming decades because of growing demand for natural gas, especially for electricity generation.

Coal is expected to remain the dominant fuel in developing Asia, especially in India and China, which have large reserves. As with other fossil fuels, coal supplies are finite, but world coal reserves are likely to be adequate for a long time to come. The chief future problem with coal is the pollution produced when it burns, which includes both toxic substances and greenhouse gases.

If fossil fuels become scarce or unacceptable for environmental reasons, other energy sources must take their place. Some analysts think that nuclear power, despite all the criticisms aimed at it, will prove the best replacement fuel. They point out that nuclear plants give off no carbon dioxide or pollutant gases. New plant designs such as the gas-cooled pebble bed modular reactor, in which small spheres of uranium (the "pebbles") heat helium gas, are said to be far safer than old ones. "If you're really concerned about global warming and carbon dioxide emissions," Vice President Dick Cheney told the 12th annual Energy Efficiency Forum in 2002, "we need to aggressively pursue the use of nuclear power."[49] Environmental groups, however, have questioned claims about the safety of nuclear energy. Corinne Veithen of Friends of the Earth in Austria said in 2001 that "solving the problem of climate change with nuclear energy is like replacing the plague with cholera."[50]

One hope for the distant future lies in a different kind of nuclear reaction: fusion, which powers the Sun. In a fusion reaction, tremendous heat and pressure force the nuclei of four hydrogen atoms (protons) together to form a helium nucleus, two positrons (particles like electrons except that they have a positive electrical charge rather than a negative one)—and a huge amount of energy. (A hydrogen bomb is an uncontrolled fusion reaction.) Unlike fission, fusion produces no radioactive waste, and its fuel, hydrogen, exists in essentially infinite supply. It produces no carbon dioxide or pollutant gases, either, and fusion reactors can be made such that the laws of physics will shut them down in case of a malfunction.

Unfortunately, as Donald L. Barlett and James B. Steele wrote in *Time* in July 2003, fusion "has been an elusive goal for half a century and probably will be for many decades to come."[51] Achieving the temperature and density of protons needed for a fusion reaction has proved almost impossible. Fusion reactions take place in a superhot, gaslike material called a plasma, which consists only of protons. Protons all have a positive electrical charge, and particles with the same kind of charge repel each other very strongly, so the plasma will fly apart unless it is somehow contained long enough for the reaction to take place. So far, a magnetic field seems the best "container" for the plasma. Fusion reactors that use magnetic fields are called tokamaks. Even the best tokamaks have not quite achieved the break-even point, where they produce more energy than is required to start them up. They are even further from the ignition stage, in which the reaction produces enough energy to sustain itself in spite of the losses that inevitably occur, including the energy taken for human use.

The future of hydropower is as much in doubt as the future of nuclear power. The EIA expects moderate growth in use of this energy source, mostly due to completion of large hydroelectric plants in developing Asia, such as China's immense Three Gorges Project. Both because most of the best sites have been taken and because environmentalists have become increasingly critical of large hydro projects, however, only Canada among developed nations is planning any. On the other hand, small projects with low-head dams have environmental groups' blessing, and former Carnegie-Mellon engineering professor Robert U. Ayres claimed in late 2001 that hydropower from such projects is "by far" the least expensive of noncarbon energy sources.

One potential future problem for hydropower is the very thing that might otherwise encourage its use: global warming. Hydroelectric power would be advantageous in a warming world because it produces no greenhouse gas emissions, but the warming may increase evaporation and change rainfall patterns, thereby reducing the amount of the river water on which hydro depends. A 2000 report by the World Commission on Dams suggested that large hydropower projects may even add to the greenhouse effect, especially in tropical climates, because their dams create large amounts of rotting vegetation, which gives off methane.

Among renewable energy sources other than hydropower, wind power is the one that the EIA expects to grow most in the next 20 years or so, especially in Western Europe and the United States. Britain, Denmark, and Germany are planning or have installed large offshore wind farms, for instance. Biomass and geothermal power are also predicted to grow rapidly in the United States. Other renewable energy sources, including solar power, are considered to be "long shots" that will not be a major part of the energy mix for decades, if ever.

There is one possible exception: hydrogen. Although everyone agrees that their widespread use is at least a decade away, hydrogen fuel cells have attracted far more excitement than any other renewable energy source in the first years of the 21st century. Extreme boosters of fuel cell technology such as Jeremy Rifkin, founder and president of the Foundation on Economic Trends and a perennial critic of mainstream technology, even claim that hydrogen could eventually become the basis for an entire "hydrogen economy"—an idea that French science fiction author Jules Verne first proposed in an 1874 book called *The Mysterious Island.* Hydrogen, Rifkin and other supporters say, not only could mean a virtual end to energy-related pollution but could provide a way of, in effect, storing electricity. In their vision, renewable, nonpolluting energy sources such as wind would be used to make hydrogen from water by electrolysis. The hydrogen would be kept in tanks and used in fuel cells to generate electricity as needed.

How much renewable energy sources as a whole will add to the world's future energy supply is a matter of hot debate. The EIA expects the contribution to be tiny, even if conventional hydropower is included. Critics of renewables say that, unless use of these energy sources is artificially supported with substantial government subsidies, they will remain uneconomical compared to fossil fuels for the foreseeable future—and perhaps always. An editorial in the January 6, 2003, issue of *Oil and Gas Journal* claimed that the idea that a combination of conservation and renewable energy sources can replace fossil fuels was "more than wishful thinking. It's dangerous fantasizing."[52]

Supporters of renewable energy, by contrast, claim that a considerable growth in the use of renewables is not only feasible but beginning to happen. "Solar power, wind turbines and other sources of clean power . . . are now poised on the brink of the mainstream," Mark Townsend wrote in *Geographical* in June 2002.[53] Even if renewable sources are not yet fully competitive in price, these advocates stress, the dwindling supplies of fossil fuels and, even more, the growing environmental damage caused by extracting and burning such fuels will make it worth paying more, if necessary, for alternative energy. A 2003 article in the *Weekly Petroleum Argus* summed up the conflict between fossil fuels and alternative or renewable resources nicely: "To the energy industry, the world cannot afford to suddenly go green. To the environmental lobby, it cannot afford not to."[54]

Some large oil companies, especially BP (which now claims that the letters in its name stand for "Beyond Petroleum") and Royal Dutch/Shell, are hedging their bets—and garnering some positive publicity at the same time—by making substantial investments in renewable energy research. BP stresses wind and solar power, while Royal Dutch/Shell specializes in solar photovoltaic, hydrogen, biomass, and geothermal energy. "The case for shifting to alternatives is gaining momentum," Heesun Wee noted in *Business Week Online* on September 25, 2002. "The game plan for companies like BP and Royal Dutch is to be ready when that changeover occurs . . . and . . . reap a financial windfall."[55] Other oil giants, such as ExxonMobil, remain much more skeptical about the future of renewables.

DOMESTIC ENERGY ISSUES

Unless perhaps pressured by rising oil and gasoline prices, supply disruptions triggered by events in the Middle East or elsewhere, or major weather changes clearly attributable to global warming, Congress seems unlikely to agree on a comprehensive energy policy any time soon. Efforts to match domestic energy supply to demand will probably continue to combine pressures for more oil drilling, subsidies for renewable energy, and increases in energy efficiency. Drilling seems likely to take place offshore and in Alaska but not in the Alaska National Wildlife Refuge. Use of renewable energy sources probably will increase as (and if) their cost drops, with or without government subsidies, but they will remain only a small part of the U.S. energy picture. If prices of gasoline and electricity stay as high or higher than they were in 2004, consumers will show increased interest in fuel-efficient technology such as gas-electric hybrid vehicles, especially if available models come to include the popular SUVs.

Pollution from motor vehicles probably will not decrease much unless fossil fuel–burning engines are replaced, an event unlikely to happen soon. Emissions from electric power plants, on the other hand, most likely will drop, at least in developed countries, as coal-powered plants are retired and replaced with gas-fired ones. Lack of investor enthusiasm caused by deregulation, difficulties in obtaining permission to build new plants, and stringent environmental requirements for such plants could slow this process. If public and political concern about global warming increases, carbon dioxide may be added to the list of pollutants limited by the Clean Air Act, or emissions of CO_2 and other greenhouse gases may be taxed or otherwise regulated. Any resulting environmental improvement would probably come at the cost of increased prices for electricity and other products, however.

Restructuring and deregulation of electricity and other energy industries, leading to increased competition in an open market, seem sure to continue in most or all of the developed world. Mindful of mistakes made in places like California, however, states and countries are likely to approach deregulation cautiously.

In the United States, the ongoing conflict between the states and FERC about who should control the flow of electricity is likely to continue. FERC is pushing for placing as few as four Regional Transmission Organizations (RTOs—Independent System Operators or transcos) in charge of the nation's entire transmission network, claiming that having a small number of large RTOs will increase efficiency and reliability as well as help prevent the sort of market manipulation that occurred during the California energy crisis. Many state governors oppose this approach, however, because they fear that such organizations would take away states' control of energy regulation within their borders.

A similar argument has erupted over a set of rules called Standard Market Design (SMD), which FERC proposed in 2002. Governors such as Jim Geringer, former governor of Wyoming, say that SMD would give FERC too much power, but other observers, including John B. Howe, former chairman of the Massachusetts PUC, claim that establishing a single set of regulations for the whole country would ease the confusion and uncertainty that keep investors from putting money into building new power plants and repairing or upgrading transmission lines.

An important question for the future of electricity deregulation is whether it should be applied to the retail as well as the wholesale market. If it is, residential customers will be able to select their electricity generators, just as they now choose long-distance telephone carriers. FERC has predicted that some degree of open-market competition will exist in retail as well as wholesale electricity sales by 2011, and some states already permit it. Analysts disagree, however, about whether households will be interested in such an option (in states where it exists, most seem not to be) and whether it will benefit them or the environment. It could allow consumers to specify "green power," for instance, but if it makes prices drop, it might harm the environment indirectly by encouraging increased consumption.

If retail electricity sales are deregulated, "smart meters" in consumers' homes may alert them to changes in price that occur as demand varies with time of day. They could then reduce their energy use at times of peak demand or even program nonessential appliances to turn off when electricity prices rise above a certain level. This not only would save them money but would even out demand, thereby cutting down on the number of power plants needed to meet peak demand and reducing strain on the electricity grid, which in turn would increase the system's reliability. According to the Electric Power Research Institute, such meters could also send second-by-second information about demand back to central computers at RTOs, allowing them to distribute loads more appropriately.

Aging power plants and transmission grids will need to be upgraded or replaced soon in many nations to avoid shortages and blackouts, most observers agree. Just after the August 2003 northeastern blackout, which demonstrated the system's weaknesses so spectacularly, *Newsweek* science and technology reporter Brad Stone wrote, "If the electricity grid is our nation's circulatory system, then America desperately needs a triple bypass."[56]

Upgrading the grid with improved technology such as high-temperature superconducting cables could increase electricity flow and reduce congestion. Many experts say that the network also needs more integrated management systems and greater automation and computerization, including instantaneous communication over wide areas. A more heavily computerized system would distribute power more efficiently, thus saving energy, and would help to avoid the sort of confusion that caused the August 2003 blackout. Cybersecurity researcher Eric Byres warns, however, that increased dependence on computers for grid management would put the system at greater risk from damage by hackers or terrorists.

Regulations governing long-distance transmission management also need to be clearer and more strictly enforced, analysts say. The U.S.-Canadian panel that investigated the August blackout recommended that the rules for transmission management developed by the North American Electric Reliability Council be written into federal law and that punishments be instigated for violators. The panel claimed that the Midwest Independent System Operator, the regional agency that it partly blamed for the blackout, broke several of these rules.

To work around private investors' unwillingness to spend money on the transmission system, Brad Stone recommends "creat[ing] a Marshall Plan for the grid," using federal funds to pay for major transmission upgrades and allowing FERC to offer incentives for power generators to build new transmission lines. Alternatively, John Howe, vice president for electric industry affairs at American Superconductor Corporation, says that the government should guarantee utilities a higher rate of return on their investments in the grid and the ability to pass on construction costs to ratepayers outside their service area, because these ratepayers also benefit from grid improvements.

Upgrading could allow the transmission network to handle input from diffuse, intermittent power sources such as wind and solar energy, which it cannot easily do at present. Upgrades could also prepare the grid to deal with a new phenomenon that may become widespread in future years: distributed generation. Ever since Thomas Edison's day—or, more precisely, George Westinghouse and Samuel Insull's—electricity transmission has been designed in a hub-and-spoke pattern, in which power flows only from central generating plants to distribution networks and, eventually, individual users. Some commentators think, or hope, that that will change, returning the system to something like Edison's original vision, in which most large and even many small users generate most or all of their own power.

Instead of consisting of hubs and spokes, a grid built for distributed generation would look more like a web. A multidirectional web of this kind is a centerpiece of the "hydrogen economy" imagined by Jeremy Rifkin and others. Deregulation laid the legal groundwork for an electrical transmission web by mandating the opening of transmission lines to all generators, and computer programs and other technology for coordinating small power sources and integrating them into a central grid are beginning to make such a structure physically possible as well. Some states already allow businesses and homes that make their own electricity to feed excess power back into the grid, and small electricity distribution networks, or microgrids, are also beginning to appear throughout the world. If distributed generation becomes widespread, today's utilities may metamorphose into "virtual utilities," primarily coordinating and packaging the flow of many small generators' electricity rather than providing power themselves.

Supporters of distributed generation point to many advantages. Having electricity generated close to the places where it is used saves the considerable expense and energy loss involved in long-distance transmission and distribution. Industries can often both save money and increase the reliability of their power supply by generating some or all of their own electricity. Small power installations can use nonpolluting energy sources such as sun and wind more easily and economically than large plants can. Small power plants are also less tempting targets for terrorists than large ones, and attacks or accidental equipment failures on a distributed generation web would be less likely to have far-reaching effects than failures on the present grid. Perhaps ultimately most important, small plants in individual villages in the developing world could leapfrog over the need to build conventional power grids and provide electricity to the billions of people worldwide who lack it.

Distributed generation encourages control of electricity supplies by communities or even, to some extent, individuals rather than the traditional investor-owned utilities. Ultimately, Jeremy Rifkin and other visionaries say, the combination of distributed generation and hydrogen as an electricity storage medium could give communities, not only power in the sense of energy but political power as well, because the communities would no longer have to depend on governments or large companies for their energy supply. "Were all individuals and communities in the world to become the producers of their own energy, the result would be a dramatic shift in the configuration of power," Rifkin writes.[57] He believes that such a shift could "fundamentally change the nature of our market, political and social institutions."[58]

Not surprisingly, established utilities have usually been less enthusiastic about distributed generation. Until recently they saw it as a nuisance at best and a competitive threat at worst, and they often tried to limit small generators' access to the grid by charging high connection fees. Some are now beginning to support distributed generation, however, because they realize it can take strain off their transmission systems and reduce the need to build new power plants. Most analysts see distributed generation and the conventional power system as coexisting in the future, though opinions vary about how much each will contribute to the total electricity picture.

Global Energy Issues

Unless the United States, and the world, frees itself from dependence on fossil fuels, changes in oil, gasoline, and natural gas prices will continue to have major effects on world economies, and the internal politics of oil-containing regions will go on being of acute concern to everyone. Assurance of continued supplies of fossil fuels—by whatever means necessary—will almost surely play a large part in the national security policies of developed countries such as the United States and large, rapidly developing nations such as China and India.

Iraq was handed over to an interim native government at the end of June 2004 and elections were held there in January 2005, but instability, sabotage of oil-related facilities, and a U.S. military presence in the country are sure to continue for years. The violent attacks on Americans in Saudi Arabia in 2004 focused increasing attention on that even more crucial country as well. In the future, Iran, parts of Africa, the Caspian Sea area, the South China Sea, and Colombia and Venezuela may join the perennial Middle Eastern "hot spots" on the front pages of newspapers worldwide. Still other locations, as yet unknown, could be added to the list if they are found to contain major deposits of oil or gas or they become sites of crucial pipelines or tanker traffic.

Multinational oil companies are sure to find the social and environmental effects of their projects, especially in developing countries, scrutinized with ever-increasing care by international lending institutions, activist groups outside the countries, and citizens of the countries themselves. This will be particularly true if any of the pending lawsuits against the companies are successful. Even if the suits fail, analysts say that companies will find it advisable to "clean up their act" if they want to stay competitive in the world market.

The watchword for international projects today is "sustainable development." As with most popular catchphrases, people disagree on what this one means. The United Nations World Commission on Environment and Development has defined it as "meet[ing] the needs of the present without compromising the ability of future generations to meet their own needs."[59] Sustainable development can include such things as respecting local customs, creating jobs and providing social services for indigenous people, using resources wisely, and protecting the environment. Most important, policy experts stress that each country—and, often, different communities and groups within a country—will have its own definition of sustainable development, and companies will need to understand and comply with those different visions if they want to gain access to the countries' energy wealth.

Managing projects that take energy out of a country is only part of the sustainable energy development story—and probably not the most important part. At the United Nations World Summit on Sustainable Energy in 2002, the International Energy Agency stressed that global sustainable development will be possible only if inexpensive electricity is made available to the world's poorest people. The IEA said that at least 1.6 billion people worldwide have no access to electricity, and 2.4 billion do all their cooking and heating with wood and agricultural or animal waste, the most primitive and polluting of fuels. Lack of electricity and better fuels perpetuates these people's poverty because it keeps them from setting up industries that would create jobs.

Far from sharing Jeremy Rifkin's vision of a future dominated by hydrogen, renewable resources, and local generation projects, the IEA felt that most new electricity for the world's urban and rural poor would come from fossil fuels and conventional grids. The agency was not hopeful about obtaining the huge investment that worldwide electrification would require, however, and predicted that even in 2030, 660 million people in sub-Saharan Africa and 680 million in southern Asia would still lack electricity.

Motorization as well as electrification is likely to be the focus of energy projects in developing countries, as more far-seeing governments try to head off the multiplication of small, highly polluting internal combustion engines. Some developing countries, especially in Latin America, are financing roads, railroads, and other transportation facilities by letting private companies operate or own them. China and other countries are promoting electric bicycles and scooters. Both Asia and Latin America are creating bus rapid transit systems that aim to produce the effect of railways at less cost. The buses in these systems have signal priority over other traffic or operate on separate roads.

The global energy issue sure to be of the greatest concern in the future, however, is the world's effort to avoid or minimize global warming by reducing emissions of carbon dioxide, methane, and other greenhouse gases. In 2004, the EIA predicted that world carbon dioxide emissions would rise from 23.9 billion metric tons in 2001 to 27.7 billion metric tons in 2025, mostly due to the increased use of fossil fuels, especially coal, in the developing world. The Intergovernmental Panel on Climate Change (IPCC), established by the United Nations and the World Meteorological Organization, has projected that, if left unchecked, increases in greenhouse gases will produce an average global temperature rise of 2.7 to 10.4 degrees F by 2100.

The World Health Organization has made the sobering prediction that warming will be hardest on the world's poor, most of whom live in the tropics. For instance, it is likely to reduce crop yields, increase the range of disease-bearing mosquitoes, and destroy ill-constructed homes in mudslides and storms. However, experts say that even nations like the United States, which possess both a more temperate climate and more resources for adapting to change, will not escape its effects. Studies have predicted that global warming will eventually cost developed countries 1 to 2 percent of their gross domestic product. In 2004 a Pentagon planning unit also warned the U.S. government that global warming was an urgent national security threat because the natural calamities it will cause could produce mass starvation and trigger wars, perhaps even nuclear ones. All but the most diehard skeptics agree that the world must try to prevent these developments.

The Kyoto Protocol is currently the chief international treaty governing the world's attempt to control global warming. In addition to ratifying the protocol, some countries or groups of countries, notably Britain and the European Union, have set their own targets for reducing greenhouse gas emissions, which are often more stringent than those mandated by Kyoto. The European Union, for instance, hopes to reduce emissions by 8 percent (relative to its 1990 levels) by 2010, and Britain has vowed to reduce its emissions by 20 percent by the same date. By contrast, Kyoto requires Britain to reduce its emissions by only 12.5 percent.

Many observers, however, doubt whether these challenging goals, or even the less strict ones of the Kyoto agreement, will or perhaps can be met. What will happen after the protocol becomes international law on February 16, 2005, if countries fail to achieve the goals it sets, is unclear. Although the treaty specifies penalties for nations that do not meet their targets, it has no mechanism for enforcement or even verification of the emission levels that countries claim to have achieved. Of course, nature may apply its own penalties in the form of climate change. Indeed, even fulfilling Kyoto's demands may not be enough to reverse or reduce global warming.

Some critics have called the Kyoto Protocol too severe, with the potential to cause major economic damage, while others have said it is too weak. Many think it should be substantially revised or even scrapped and replaced with something quite different. Several analysts say it should cover developing as well as developed countries and should restrict emission of all greenhouse gases from all sources rather than focusing on carbon dioxide and the burning of fossil fuels. Some favor an expanded cap-and-trade system, while others prefer international emissions taxes. Commentators including Tom Blundell, chair of Britain's Royal Commission on Environmental Pollution, say that emission targets should eventually be assigned to countries on a per capita basis.

A few people call for more severe measures still—nothing less than a restructuring of world society. For example, Mayer Hillman, senior fellow emeritus of the Policy Studies Institute in London, says that controlling greenhouse gas emissions sufficiently to stop serious and possibly irreversible climate change will require giving up not only fossil fuels but the entire goal of economic growth. Policies favoring growth, he claims, "must be replaced by policies focused on improving quality of life against the background imperatives of delivering sustainability and equity." He admits that "the measures needed to promote a substantial modification of our lifestyles will necessarily be draconian."[60]

Hillman's words echo a more positive vision that M. King Hubbert, the renowned geophysicist, is said to have espoused. According to Richard Heinberg, writing in the Autumn 2003 *Earth Island Journal*, Hubbert believed that "all we need to do is overhaul our culture and find an alternative to money. If society were to develop solar-energy technologies, reduce its population and its demands on resources, and develop a steady-state economy to replace the present one based on unending growth, our species' future could be rosy indeed."[61] Neither Hillman nor Hubbert explained how these eye-opening changes might be brought about. Most likely they would require the imminent threat of worldwide catastrophe—and by then, societal changes, no matter how sweeping, might come too late to head off disaster.

---

[1] E. F. Schumacher, quoted in Wayne Ellwood, "Mired in Crude," *New Internationalist*, June 2001, p. 10.

[2] General Accounting Office, quoted in Jerry Taylor and Peter VanDoren, "Evaluating the Case for Renewable Energy," *Power Economics*, vol. 6, March 2002, p. 25.

[3] Lita Epstein, C. D. Jaco, and Julianne C. Iwersen-Neimann, *The Complete Idiot's Guide to the Politics of Oil.* New York: Penguin Group/Alpha, 2003, p. 4.

[4] Henry Linden, "Energy Independence Now," *Public Utilities Fortnightly*, vol. 141, July 1, 2003, p. 22.
[5] Lewis Strauss, quoted in Marek Walisiewicz, *Alternative Energy*. New York: DK Publishing, 2002, p. 22.
[6] *Forbes*, quoted in Walisiewicz, *Alternative Energy*, p. 23.
[7] Fred Bosselman, Jim Rossi, and Jacqueline Lang Weaver, *Energy, Economics and the Environment: Cases and Materials*. New York: Foundation Press, 2000, pp. 956–957.
[8] Kate Tsubata, "Nuclear Power's New Promise and Peril," *World & I*, vol. 17, May 2002, n.p.
[9] Diane D'Arrigo, quoted in Karl Grossman, "The Nuclear Phoenix," *E*, vol. 12, November–December 2001, p. 36.
[10] "Bill Summary & Status for the 94th Congress, S.622, Public Law: 94-163" (which became the Energy Policy and Conservation Act).
[11] George W. Bush, quoted in "Strategic Deterrent," *Global Markets*, vol. 33, May 5, 2003, p. 1.
[12] *New York Times* editorial, July 14, 1994, quoted in Bosselman, Rossi, and Weaver, *Energy, Economics and the Environment*, p. 393.
[13] Gale Norton, quoted in Zachary Coile, "Senate Blocks Bush's Oil Drilling Plan for Alaska Refuge," *San Francisco Chronicle*, March 20, 2003, p. A16.
[14] Taylor and VanDoren, "Evaluating the Case for Renewable Energy," p. 25.
[15] Gregg Easterbrook, "The Producers—How the Oilmen in the White House See the World," *The New Republic*, June 4, 2001, p. 28.
[16] Michael Economides, "The Future of Energy," *Offshore*, vol. 62, June 2002, p. 108.
[17] Mortimer B. Zuckerman, "Speaking Truth About Energy," *U.S. News & World Report*, February 18, 2002, p. 68.
[18] Gray Davis, quoted in Roger James, "Lessons from California," *Petroleum Economist*, vol. 69, February 2002, p. 3.
[19] Kevin and Bob (Enron traders), quoted in Zachary Coile, "New Evidence of Enron Schemes," *San Francisco Chronicle*, June 15, 2004, pp. A1, A11.
[20] Steven N. Isser, "Electricity Deregulation: Kilowatts for Nothing and Your BTUs for Free," *Review of Policy Research*, vol. 20, Summer 2003, p. 238.
[21] Matthew L. Wald, Richard Pérez-Peña, and Neela Banerjee, "What Went Wrong," *New York Times*, August 16, 2003, p. A1.
[22] Philip E. Clapp, quoted in John Carey, "Taming the Oil Beast," *Business Week*, February 24, 2003, p. 96.
[23] Sarah Searight, "Region of Eternal Fire," *History Today*, vol. 50, August 2000, pp. 45ff.
[24] Ian Gary, quoted in "Bottom of the Barrel," *Energy*, vol. 28, Summer 2003, p. 32.
[25] Ellwood, "Mired in Crude," pp. 9ff.
[26] Epstein, Jaco, and Iwersen-Neimann, *The Complete Idiot's Guide to the Politics of Oil*, p. 187.
[27] Ellwood, "Mired in Crude," pp. 9ff.
[28] Donald Losman, "Oil Is Not a National Security Issue," *USA Today*, vol. 130, January 2002, p. 19.

[29] Lord George Nathaniel Curzon, quoted in Michael T. Klare, *Resource Wars*. New York: Henry Holt, 2002, p. 30.
[30] Josef Stalin, quoted in Kevin Adler, "Oil Refining Changes Transportation, History, and Ways of Life," *Chemical Week*, vol. 164, September 18, 2002, pp. SS30ff.
[31] Jimmy Carter, quoted in Klare, *Resource Wars*, p. 4.
[32] Daniel Yergin, "Gulf Oil—How Important Is It, Anyway?" *San Francisco Chronicle*, April 13, 2003, p. E1.
[33] Michael Economides and Ronald Oligney, "The Color of Oil (2)," *Energy*, vol. 25, Fall 2000, pp. 17ff.
[34] Amory B. Lovins and L. Hunter Lovins, "Supply-Side Stupor," *The American Prospect*, vol. 13, January 28, 2002, p. 21.
[35] Donald Rumsfeld, quoted in Jim Vallette, Steve Kretzmann, and Daphne Wysham, "Crude Vision: How Oil Interests Obscured U.S. Focus on Chemical Weapons Use by Saddam Hussein." Sustainable Energy and Economy Network Institute for Policy Studies. Available online. URL: http://www.seen.org/PDFs/Crude_Vision2.pdf. Posted August 13, 2002. p. 2.
[36] J. Robinson West, "Five Myths About the Oil Industry," *International Economy*, vol. 17, Summer 2003, p. 45.
[37] Michael T. Klare, "Oil Moves the War Machine," *The Progressive*, vol. 66, June 2002, p. 19.
[38] Ed Ayres, "'It's Not About Oil,'" *World Watch*, vol. 16, May–June 2003, p. 4.
[39] John Bacher, "Petrotyranny," *Earth Island Journal*, vol. 17, Spring 2002, p. 34.
[40] Ellwood, "Mired in Crude," pp. 9ff.
[41] Daniel Fisher, "Dangerous Liaisons," *Forbes*, vol. 171, April 28, 2003, p. 84.
[42] Kenny Bruno, "De-globalizing Justice," *Multinational Monitor*, vol. 24, March 2003, p. 14.
[43] International Chamber of Commerce, quoted in Bruno, "De-globalizing Justice," p. 14.
[44] Daniel Griswold, quoted in Bruno, "De-globalizing Justice," p. 14.
[45] Burmese rice farmer, quoted in Adam Zagorin, "Slave Labor?" *Time*, vol. 162, November 24, 2003, p. A18.
[46] William Jobin, "Health and Equity Impacts of a Large Oil Project in Africa," *Bulletin of the World Health Organization*, vol. 81, June 2003, p. 420.
[47] Framework Convention on Climate Change, quoted in Stephen M. Gardiner, "The Global Warming Tragedy and the Dangerous Illusion of the Kyoto Protocol," *Ethics and International Affairs*, vol. 18, April 2004, p. 23.
[48] Robert Kaufmann, quoted in Tim Appenzeller, "The End of Cheap Oil," *National Geographic*, vol. 205, June 2004, p. 108.
[49] Dick Cheney, quoted in Tsubata, "Nuclear Power's New Promise and Peril," n.p.
[50] Corinne Veithen, quoted in Christine Laurent, "Beating Global Warming with Nuclear Power?" *UNESCO Courier*, February 2001, pp. 37ff.
[51] Donald L. Barlett and James B. Steele, "The U.S. Is Running Out of Energy," *Time*, vol. 162, July 21, 2003, pp. 36ff.
[52] "A Wishful-Thinking Wish," *Oil and Gas Journal*, vol. 101, January 6, 2003, p. 17.
[53] Mark Townsend, "Power to the People," *Geographical*, vol. 74, June 2002, p. 58.

[54] "Warming Relations," *Weekly Petroleum Argus*, vol. 33, August 11, 2003, p. 1.
[55] Heesun Wee, "Can Oil Giants and Green Energy Mix?" *Business Week Online*, September 25, 2002, n.p.
[56] Brad Stone, "How to Fix the Grid," *Newsweek*, August 25, 2003, p. 38.
[57] Jeremy Rifkin, "Hydrogen: Empowering the People," *The Nation*, vol. 275, December 23, 2002, pp. 20ff.
[58] Rifkin, "Hydrogen: Empowering the People," p. 20.
[59] The United Nations World Commission on Environment and Development, quoted in Jim Miller, "Sustainable Development," *Oil and Gas Investor*, vol. 23, August 2003, p. 44.
[60] Mayer Hillman, "The Options Are Awesome," *Energy and Environmental Management*, January–February 2003, p. 16.
[61] Richard Heinberg, "The End of the Oil Age," *Earth Island Journal*, vol. 18, Autumn 2003, pp. 24ff.

CHAPTER 2

THE LAW AND ENERGY

LAWS AND REGULATIONS

Numerous federal, state, and local laws affect the generation, distribution, and use of energy in the United States. So do hundreds of regulations by the Federal Energy Regulatory Commission (FERC), state public utilities commissions, and other government agencies. This section describes the federal laws that have had the most significant effects on the energy issues described in chapter 1. The laws are arranged by date, with the oldest first.

PUBLIC UTILITY HOLDING COMPANY ACT (PUHCA, 1935)

The utilities that distributed electricity and natural gas began as small, locally managed operations. By 1930, however, 90 percent of the electricity generated in the United States came from utilities owned by holding companies, which were businesses formed for the purpose of buying controlling shares of stock in other businesses. In many cases, the holding companies were owned by still other holding companies, forming a many-layered pyramid of businesses related in complex ways, with only a handful of companies at the top. The utilities that actually produced and distributed gas and electricity were at the bottom. Holding companies abused the utilities and their ratepayers by, for example, charging exorbitant rates for loans and technical assistance.

Utilities were traditionally regulated at the local or state level, but since the holding companies owned utilities in many states and sent gas and electricity from one state to another, such regulation did not affect them. When the holding companies' abuses became obvious, the public demanded that the federal government take a hand. The result was the Public Utility Holding Companies Act (PUHCA, 15 U.S.C. 79–79z), which President Franklin D. Roosevelt, who strongly supported it, signed into law on August 26, 1935. This law was part of a larger bill called the Public Utility Act, which also revised the structure of the Federal Power Commission, the ancestor of today's FERC, and gave it control of interstate transmission and wholesale sales of electricity.

According to a booklet put out by the Department of Energy (DOE), PUHCA "shaped the electric industry for over half a century."[1] It required utility holding companies to register with the Securities and Exchange Commission, which was put in charge of administering and enforcing the law, and provide the commission with detailed accounting information about their activities. In most cases, it forced the companies to limit their operations to physically and economically interconnected systems in particular regions, often consisting of single states, and forbade companies to own both utilities and nonutility businesses.

About a quarter of investor-owned utilities are still subsidiaries of utility holding companies and, as such, are regulated by PUHCA. The Energy Policy Act (EPAct) of 1992, however, revised PUHCA to allow utilities to own independent power producers and to exempt some wholesale generators of electricity from the earlier law.

THE CLEAN AIR ACT (1970)

The Clean Air Act (CAA, 42 U.S.C. 7401–7642) is the chief U.S. law that limits air pollution, including pollution from electric power plants and automobiles. These two sources account for most of the air pollution in the United States.

The Clean Air Act was first passed as the Air Pollution Control Act in 1955, which merely provided for "research and technical assistance" on the subject. The law was put into its present basic form in 1970 but substantially revised in 1977, 1990, and 1997. The Environmental Protection Agency (EPA) oversees implementation of this law, but the states devise their own plans and standards for carrying it out and submit them to the EPA for approval. The state plans can be stronger, but not weaker, than the overall CAA. The EPA also sets up interstate commissions to develop regional strategies to control air pollution that moves from one state to another or between the United States, Canada, and Mexico.

Under the CAA, the EPA sets countrywide limits (National Ambient Air Quality Standards) on the emission of six types of pollutants. These pollutants, called criteria air pollutants, are believed to cause significant harm to health, the environment, or property. They include ground-level ozone, sulfur dioxide (SO_2), nitrogen oxides, carbon monoxide, particulates, and volatile organic compounds. Areas that meet or exceed the standards for criteria pollutants are called attainment areas, whereas those that do not meet the standards are nonattainment areas. Many urban areas are nonattainment with regard to at least one pollutant.

The version of the Clean Air Act passed in 1970 required inspection of new or remodeled coal-burning power plants to ensure that they used the "best available control technology" to limit emission of sulfur dioxide, a process called new source review. (The 1997 revision of the act added other criteria pollutants as well.) Plants already in existence and not remodeled, however, were exempted from these standards. This provision has been attacked by both environmentalists and members of the power industry, who agree that it encourages plant owners to keep old, highly polluting plants rather than invest in new or remodeled ones. In 2003, the George W. Bush administration waived the new source review requirement for remodeled plants.

In addition to the new source review provision, the Clean Air Act limits sulfur dioxide emission from power plants through a market-based mechanism called a cap-and-trade program. This program, established in Title IV of the act's 1990 revision, sets an upper limit, or cap, for the total amount of SO_2 emitted in the country and assigns each plant a certain number of "emission allowances" as a way of specifying the amount of SO_2 it may give off without penalty. It can choose whatever technology it wishes to achieve this level. Companies that reduce their pollutants below their assigned amounts can sell their unused emission allowances to other businesses that are unable or unwilling to make similar reductions. This approach, supporters say, produces an overall lowering of pollutants but allows the market to determine the most efficient mechanism for achieving this effect. The act establishes other cap-and-trade programs to produce cleaner-burning gasoline and reduce emission of chemicals classified as "air toxics."

The Clean Air Act requires periodic tests of "mobile sources"—that is, motor vehicles—to make sure that they are maintained in ways that keep their pollutant emissions low. One provision of the act also affects the composition of the gasoline and diesel fuel that vehicles use. This provision was first used to phase out the sale of leaded gasoline starting in 1975. The 1990 revision of the act required refiners to add oxygenating (oxygen-containing) compounds to gasoline sold in certain states during the winter to reduce carbon monoxide pollution, as well as setting limits on the amount of sulfur in diesel fuel and, later, gasoline.

ENERGY POLICY AND CONSERVATION ACT (1975)

The Energy Policy and Conservation Act (P.L. 94–163), passed in the wake of the 1973–74 Arab oil embargo, authorized the president, with congressional approval, to take a variety of measures to protect the country from future shortages of imported oil and encourage energy conservation. It became law on December 22, 1975.

Steps taken by this act to ensure adequate petroleum supplies included establishment of the Strategic Petroleum Reserve (SPR), a stockpile of up to 1 billion barrels (the actual maximum capacity as of 2004 is 700 million barrels) of domestic oil that could be called upon, or "drawn down," when "the President has found that . . . [it] is required by a severe energy supply interruption." The act also provided economic incentives for increasing domestic production of crude oil and other energy sources, including drilling in waters off U.S. shores and establishing new coal mines.

On the conservation side, the act established Corporate Average Fuel Efficiency (CAFE) standards, which set minimum requirements for motor vehicle fuel efficiency, measured in miles per gallon. The requirements were different for different classes of vehicle (passenger cars and light trucks) and were to be averaged over each automaker's entire fleet (that is, all models manufactured) for each type of vehicle. In other words, some models could be less efficient than the standard, as long as others were more efficient. The first CAFE passenger car requirement was 18.0 miles per gallon, beginning with the 1978 model year. This was increased to 27.5 mpg, beginning with the 1985 model year, a figure that still applies. The 1985 standard for light trucks was set at 19.5 miles per gallon and later raised to 20.5 mpg. Both figures were a substantial improvement over the average for vehicles at the time the law was passed, which was only 14 mpg for passenger cars and and 10.5 mpg for trucks.

The Energy Policy and Conservation Act also authorized the president to set up mandatory energy conservation contingency plans and a rationing program for oil, gasoline, and other petroleum products. However, these programs were never put into action.

FOREIGN CORRUPT PRACTICES ACT (1977)

The Foreign Corrupt Practices Act (FCPA, 15 U.S.C. 78dd-1-3 and 78m) was passed in the wake of a Securities and Exchange Commission investigation growing out of the Watergate scandal. This investigation revealed that more than 400 U.S. companies, including Gulf Oil, had made questionable or illegal payments to foreign politicians and officials, amounting to more than $300 million. The act prohibits making such payments for the purpose of obtaining or keeping business. The law also requires companies doing business overseas to make and keep records that accurately reflect their transactions. It covers foreign firms doing business in the United States as well as U.S. businesses acting abroad. Oil companies have often been accused of bribing officials to further their projects in developing countries, and several have been prosecuted under the FCPA.

The Department of Justice is the FCPA's chief enforcer, although the Securities and Exchange Commission also plays a role. Because actions that violate the FCPA, and prosecution for such actions, can have foreign policy implications, these two agencies may consult the State Department before proceeding. Alleged violations of the FCPA can result in criminal charges, a civil action resulting in a fine, or both. They may also produce charges of violating other laws, such as the Racketeer Influenced and Corrupt Organizations Act (RICO).

The United States was the first country in the world to make bribery of foreign officials illegal. In 1997, after almost 10 years of urging by U.S. representatives, who feared that being the only country with such a law put the United States at an economic disadvantage, 34 members of the Organization of Economic Cooperation and Development, which includes most of the world's developed nations, signed a Convention on Combating Bribery of Foreign Public Officials in International Business Transactions, which is similar to the FCPA.

PUBLIC UTILITY REGULATORY POLICIES ACT (PURPA, 1978)

The Public Utility Regulatory Policies Act (PURPA, Pub. L. 95–617, 92 Stat. 3117) was part of President Jimmy Carter's massive National Energy Act, which was intended to increase the security of the U.S. energy supply by encouraging conservation and broadening diversity of energy sources. PURPA was passed in 1978 and took effect in 1980.

PURPA authorized the Federal Energy Regulatory Commission (FERC), which administers the law, to require utilities to buy power from certain "qualifying facilities," or QFs, specifically industries that produced power through cogeneration (using waste heat from factory processes to heat steam that turned turbines to produce electricity) or small power generators (less than 80 megawatts) that used renewable energy sources such as wind or solar power. The utilities were to purchase this power at their "avoided cost," that is, the price they would have had to pay for producing it themselves. PURPA also exempted QFs from regulation under the 1935 anti–holding company law, PUHCA, which could have been a problem for cogenerators because it barred companies from owning both utility (electricity generation) and nonutility businesses, which cogenerators normally did.

Finally, PURPA gave FERC the authority to order companies that owned interstate transmission lines to carry, or wheel, power from generators they did not own, as long as doing so did not present "an undue burden." FERC seldom took advantage of this feature of the law at the time, however.

PURPA is generally held to have succeeded in its stated aim of increasing efficiency and diversity of energy sources. Furthermore, although it had not been intended to increase competition in the electricity industry, it had that effect because it revealed that numerous nonutility generators were both willing and able to enter the electricity market. This showed that electricity generation was not a natural monopoly, as had been thought. PURPA is credited with beginning the process of electricity deregulation and restructuring that continues to this day. Fred Bosselman and the other authors of *Energy, Economics and the Environment* say that PURPA's QF program "jump-started the growth of an independent energy sector."[2]

ENERGY POLICY ACT (EPACT, 1992)

The 1992 Energy Policy Act (42 U.S.C. 6801–6892), known as EPAct, is the country's most recent comprehensive set of laws governing energy. Part of EPAct deals with electricity regulation. This section expanded the process of opening parts of the industry to competition which PURPA had begun. PURPA, for instance, had authorized FERC to require some wheeling of wholesale electricity, but that provision was little used. EPAct made the wheeling authorization clearly apply to all utilities with interstate transmission lines.

Similarly, PURPA had exempted qualifying facilities from regulation as utilities under PUHCA, and EPAct extended this exemption to a broader group of companies that were exclusively in the business of wholesale electricity generation. These exempt wholesale generators (EWGs), as they were called, could build generating plants, whether these met the definition of QFs or not, and FERC could require utilities that owned transmission lines to carry their power. This change helped to create an open, competitive market in electricity generation.

COURT CASES

A number of court cases, including some that reached the Supreme Court, have affected judicial views of regulation of the energy industry and of the behavior of energy companies. They have also brought up issues that reach beyond the subject of energy, including definitions of acceptable business practice, responsibility for human rights violations, and the separation of governmental powers. The remainder of this chapter discusses some key cases in this field.

MUNN V. ILLINOIS, 94 U.S. 113 (1876)

Background

Ira Munn and George Scott owned and operated Northwestern Elevator, a grain warehouse, in Chicago beginning in 1862. In June 1872, the State of

Illinois charged their company, Munn & Scott, with violating the Warehouse Act, a state law passed in 1871. According to this law, the Northwestern Elevator was classified as a "public warehouse." The company that owned it therefore was required to obtain a license from the local circuit court, post a bond, and charge rates for storage and handling of grain that did not exceed specified amounts. The state claimed that Munn & Scott had done none of these things.

Munn and Scott pleaded not guilty but were convicted and fined $100. They appealed the case to the Illinois Supreme Court, which upheld the judgment of the lower court, and then to the U.S. Supreme Court, which heard and decided the case in the October term of 1876.

Legal Issues

Until the Civil War, states usually did not limit the fees that businesses could charge. As certain industries, especially railroads, began to grow, however, they sometimes took advantage of their legal or *de facto* monopoly power to set exorbitant rates, causing great public outcry. When states began regulating such businesses to prevent these abuses, the businesses turned to the courts, protesting that states had no right to impose laws that, in effect, deprived business owners of their private property.

Munn and Scott maintained that the Illinois Warehouse Act violated the Fourteenth Amendment to the Constitution because it deprived them of the use of their property without due process of law and denied them equal protection under the law. Attorneys for the state, however, held that the company was "engaged in a public employment, as distinguished from ordinary business pursuits," and was thus in essence a common carrier, providing an essential service and protected from competition. They said that a long history of English and American common law gave states the right to regulate businesses of this kind to prevent misuse of their monopoly power.

Decision

Morrison R. Waite, chief justice of the United States, delivered the court's majority opinion. He began by reviewing the history of laws in England and the United States that traditionally limited the prices that such businesses as ferries, mills, and inns could charge. Laws of this kind, he said, had never been found to violate the right to own and use private property. They were permitted because such businesses were "clothed with a public interest" and "when . . . one devotes his property to a use in which the public has an interest, he, in effect, grants to the public an interest in that use, and must submit to be controlled by the public for the common good." He cited cases indicating that warehouses that accepted goods from a variety of shippers, rather than serving only a single business, were included in the category of public businesses. Waite also quoted an English jurist, Lord Edward Law Ellenborough, who said that government regulation of a business, including limitations on prices charged, was appropriate when the business was a monopoly. Because of these considerations, Waite upheld the lower court's ruling against Munn & Scott.

Justices Stephen J. Field and William Strong dissented from Waite's majority opinion. Justice Field, who wrote the dissenting opinion, claimed that laws like the Warehouse Act prevented full use of private property and thus violated the Constitution, just as if the property had been seized without compensation. He questioned the concept that some businesses, including that of Munn & Scott, should be subject to special regulation because of their supposedly "public" nature. "The defendants were no more public warehousemen . . . than the merchant who sells his merchandise to the public is a public merchant, or the blacksmith who shoes horses for the public is a public blacksmith," he wrote. He said that the jurists cited by Justice Waite were referring to businesses that had been granted monopolies or other special privileges by the government, which Munn & Scott had not, and that regulation was appropriate only when such privileges had been granted.

Impact

Electric and gas utilities did not exist at the time *Munn v. Illinois* was decided. Nonetheless, this decision, along with those in several other cases involving railroads, was later used as justification for state and federal regulation of such utilities. Because energy utilities provided services on which the public depended and were usually granted monopolies in particular areas, they, like Munn & Scott's grain warehouse, were considered to be businesses "clothed with a public interest" and therefore subject to regulations aimed at protecting the public from overcharging and ensuring that the businesses performed their duties.

Standard Oil Company of New Jersey v. United States, 221 U.S. 1 (1911)

Background

American businessman John D. Rockefeller and several partners founded the Standard Oil Company of Ohio in 1870. Even before then, Rockefeller had begun his lifelong work of buying or otherwise acquiring control of not only oil wells, refineries, and coal companies, but all other businesses that affected the petroleum industry, from barrelmakers to pipelines and railroads. By taking over or making special arrangements with these businesses, such as contracts with railroads that gave his company rebates or preferential rates, Rockefeller reduced Standard's costs tremendously and therefore could underbid his competitors, eventually forcing them to sell their businesses to him or go bankrupt. These and other even less savory practices, including sabotage and strong-arm tactics, made Rockefeller famous, or infamous—not to mention very rich.

By 1882, according to a later summary by Edward D. White, chief justice of the United States, Rockefeller's empire of openly and secretly interconnected businesses, controlled by a Standard Oil Trust founded in that year, had "obtained a complete mastery over the oil industry, controlling 90 percent of the business of producing, shipping, refining, and selling petroleum and its products." As a result, Standard Oil "was able to fix the price of crude and refined petroleum, and to restrain and monopolize all interstate commerce in these products." Outcries against Rockefeller's abuses made the federal government order the breakup of the Standard Oil Trust into a number of separate companies in 1892. Nonetheless, the Rockefeller empire remained all-powerful in the petroleum industry because it still secretly controlled the supposedly separate companies through the Standard Oil Company of New Jersey, to which it had transferred the companies' stocks in 1899.

Legal Issues

Tycoons in the railroad and certain other industries headed trust empires just as all-controlling and abusive as Rockefeller's. Protests against their business practices led Congress to pass what became known as the Sherman Antitrust Act (named after Senator John Sherman, who proposed it) in 1890. The first section of the act outlawed "every contract, combination in the form of trust or otherwise, or conspiracy, in restraint of trade or commerce among the several states or with foreign nations," and the second section similarly forbade anyone to "monopolize, or attempt to monopolize, or combine or conspire with any other person or persons to monopolize, any part of the trade or commerce among the several states, or with foreign nations."

In November 1906, at the request of President Theodore Roosevelt, the Justice Department charged the Standard Oil Company of New Jersey and 33 other companies, along with Rockefeller and six other individual defendants, with conspiring "to restrain the trade and commerce in petroleum, commonly called 'crude oil,' and in the other products of petroleum, . . . and to monopolize the said commerce," thus violating the first two sections of the antitrust act. The prosecutors asked the court not only to fine the defendants as the law required but to block them from any further restraint of trade and force the Standard Oil Company of New Jersey to give up all control of the companies it owned.

Standard Oil's lawyers admitted acquiring the businesses and forming the trust but "den[ied] all the allegations respecting combinations or conspiracies to restrain or monopolize the oil trade." Rather, according to Chief Justice White's summary, they claimed that the company's success was "the result of lawful competitive methods, guided by economic genius of the highest order" and served to "stimulate and increase production, to widely extend the distribution of the products of petroleum at a cost largely below that which would have otherwise prevailed, thus proving to be . . . a benefaction to the general public."

A Missouri circuit court consisting of four judges heard the case on April 5–10, 1909, and found most of the defendants guilty of restraining trade by combining the stocks of numerous companies under the control of the Standard Oil Company of New Jersey. It ordered the companies to separate themselves from Standard. Rockefeller and his codefendants, however, appealed to the U.S. Supreme Court. The Court heard the case, which involved what White called "a jungle of conflicting testimony" amounting to some 12,000 pages, on March 14–16, 1910, and again on January 12–17, 1911. At issue in the high court's decision was not only whether Standard Oil had violated the Sherman act but also judicial interpretation of the law itself.

Decision

Chief Justice White delivered the Court's majority opinion on May 15, 1911. After summarizing the actions by which the defendants were alleged to have restrained trade in the petroleum industry and reaped "enormous and unreasonable profits," as well as the defendants' reply, he turned to interpretation of the relevant sections of the Sherman Antitrust Act.

White stressed that the first section of the act did not restrict the making or enforcing of any type of contract in itself, but rather banned only contracts that, in his words, "unduly" restrained trade. He believed that judges should use the "standard of reason," which he did not define, to determine whether a contract fit this description. Similarly, in discussing the definition of the second section, White concluded that "monopolize" did not mean the formation of a monopoly in itself, but rather only a monopoly that resulted in undue restraint of trade, a distinction also to be determined by "the rule of reason."

Turning to the facts of the case, White said that evidence established "beyond dispute" the "vast amount of property and the possibilities of far-reaching control" involved in the organization of the Standard Oil Company of New Jersey, which resulted in the trade of numerous large companies being "all managed as that of a single person." He added, "We think no disinterested mind can survey the period in question without . . . conclu[ding] that the very genius for commercial development and organization . . . manifested from the beginning soon begot an intent and purpose to . . . drive others from the field and to exclude them from their right to trade, and thus accomplish the mastery which was the end in view." White upheld the lower court's ruling that the company violated both sections of the Sherman Act and its order to break up the holdings of Standard Oil of New Jersey.

Justice John Marshall Harlan concurred in the view that Standard Oil of New Jersey had violated both sections of the antitrust act, but he disagreed strongly with White's insertion of "a standard of reason" to determine whether a contract "unduly" restrained trade. The law itself contained no such qualifiers, he stressed. He warned that the addition of this vague standard would not only lead to confusion in interpreting the law but amounted to "judicial legislation," a violation of the separation of powers required by the Constitution.

Impact

The Court's decision resulted in the final breakup of the Standard Oil empire, producing separate companies that became the ancestors of most multinational oil companies currently based in the United States. (For the record, Rockefeller kept his fortune.) Congress modified the Sherman Antitrust Act in 1914 to specifically forbid many of the actions revealed in the Standard Oil case.

OTTER TAIL POWER CO. V. UNITED STATES, 410 U.S. 366 (1973)

Background

Otter Tail Power Company, an electric utility company, sold electricity at retail in 465 towns in North Dakota, South Dakota, and Minnesota under exclusive contracts with the individual towns that generally lasted 10 to 20 years. It supplied power to 91 percent of the towns in its service area.

Between 1945 and 1970, four towns that had contracts with Otter Tail decided to establish municipal electric utilities when the contracts expired. The municipal systems could function only by buying their power at wholesale and having it transmitted to the towns, and the only available transmission lines belonged to Otter Tail. Otter Tail refused to sell wholesale power to the towns or allow them to use the company's transmission lines to carry power purchased elsewhere. It also used provisions in its transmission contract with the Federal Bureau of Reclamation to keep the bureau from selling power to the towns, and it filed lawsuits against two of the towns. All these actions helped to prevent or delay the development of the municipal systems.

Legal Issues

Attorneys for the United States brought a civil suit against Otter Tail for violation of Section 2 of the Sherman Antitrust Act, which forbids use of monopolies to restrain trade. A Minnesota district court found Otter Tail guilty, claiming that the company's only reason for blocking the sale and transmission of power to the municipal systems was its desire to prevent competition for retail electricity sales in the area. The court ordered Otter Tail to stop all its actions against existing or future municipal systems and allow power to be transmitted, or "wheeled," on its lines to any system that requested it.

Otter Tail appealed the ruling, and the U.S. Supreme Court eventually heard the case on December 5, 1972. The utility claimed that the antitrust law did not apply to it because it was governed instead by the Federal Power Act, and industries that were already regulated could be exempted from further regulation by the antitrust laws. It also claimed that the district court had no authority to order it to wheel power because that authority belonged to the Federal Power Commission (FPC), which regulated interstate commerce in electricity. The high court's ruling was expected to establish whether electric utilities were subject to the Sherman act and also what agencies, and to what extent, could require wheeling.

Decision

The Court reached its decision on February 22, 1973. Justice William O. Douglas delivered the majority opinion. He agreed with the district court that "the record makes abundantly clear that Otter Tail used its monopoly power in the towns in its service area to foreclose competition [from municipal systems] or gain a competitive advantage, or to destroy a competitor, all in violation of the antitrust laws." There were no physical reasons why Otter Tail could not sell or wheel power to the municipal systems, he wrote, so its refusal to do those things was "solely to prevent municipal power systems from eroding its monopolistic position."

Douglas ruled that the Federal Power Act did not shield Otter Tail or other electric utilities from antitrust regulation. "There is nothing in the legislative history [of the Sherman act] which reveals a purpose to insulate electric power companies from the operation of the antitrust laws," he wrote. FPC and state public utility commission regulation of utilities was not primarily concerned with possible antitrust violations, so it did not substitute for regulation under the Sherman act.

Douglas also held that the district court's ordering Otter Tail to wheel power did not usurp the authority of the Federal Power Commission because the FPC did not have the power to order wheeling. He said that the Federal Power Act had originally included a provision that would have allowed the agency to do so, but that provision had been deleted to preserve "the voluntary action of the utilities."

Justice Potter Stewart filed a partly dissenting opinion, in which the chief justice, Warren Berger, and Justice William Rehnquist concurred. Stewart claimed that the district court's ruling and Douglas's affirmation of it were "a misapplication of the Sherman Act to a highly regulated, natural-monopoly industry wholly different from those that have given rise to ordinary antitrust principles." He believed that Otter Tail's refusal to wholesale or wheel power to the municipal systems was not an antitrust violation but, rather, an expression of a legitimate and permitted business interest. The fact that Congress had decided not to include authority to require wheeling in the Federal Power Act, he said, "indicate[d] a clear congressional purpose to allow electric utilities to decide for themselves whether to wheel or sell [electricity] at wholesale as they see fit." Stewart maintained that judgment about the validity of Otter Tail's actions should have been left to the Federal Power Commission.

Impact

Before this case was finally settled, two of the four towns involved found other sources of transmission, and a third abandoned its plan for a municipal power system and instead renewed its contract with Otter Tail. Otter Tail was required to wheel power to the remaining town, Elbow Lake, Minnesota, and eventually did so.

Otter Tail Power Co. v. United States showed that utilities were subject to antitrust regulation. It also showed that under some circumstances, an electric utility that owned transmission lines could be ordered to wheel power from another generator through those lines, a necessary prerequisite for establishing competition in electricity generation. The Federal Energy Regulatory Commission, the descendant of the Federal Power Commission, later gained some authority to order wholesale electricity wheeling through the Public Utility Regulatory Policy Act (PURPA) of 1978. The Energy Policy Act (EPAct), passed in 1992, made this authority broader and more explicit by stating that all utilities with interstate transmission must wheel wholesale electricity unless doing so would compromise reliability or would not be in the public interest.

The Law and Energy

Doe v. Unocal, BC 237980 (1996)

Background

In the early 1990s, a group of four companies joined in a project to extract natural gas from the Yadana field, located off the coast of Myanmar (Burma) in the Andaman Sea, and build a pipeline from the field to Thailand, where most of the gas would be sold. The companies included French oil and gas giant TotalFinaElf, which had a 31.24 percent interest in the Moattama Gas Transportation Company (MGTC), the business formed to carry out the project, and Unocal, a California-based oil company that joined the project in 1993 through a subsidiary and obtained a 28.26 percent interest. Thai and Burmese companies owned smaller shares of MGTC.

Although most of the Yadana pipeline was offshore, the last 40 miles (65 kilometers) crossed the Tenasserim region of southern Myanmar. Construction of the pipeline began in 1996 and was completed two years later. During this time, the Myanmar government's military forces were allegedly in charge of project security and committed a number of atrocities on the Tenasserim native peoples, including forced ("slave") labor, forced relocation, murder, torture, theft, and rape.

Two groups filed separate suits against Unocal in a federal district court in Los Angeles, California, in 1996, claiming that the company was indirectly responsible for the Myanmar soldiers' actions because it knew, or should have known, about the abuses and took no steps to prevent them. The first suit, *Doe v. Unocal*, was a class action suit filed on behalf of all residents of the Tenasserim region. The National Coalition Government of Burma and the Federation of Trade Unions of Burma filed the second suit, *Roe v. Unocal.* Human rights groups, including Earthrights International (for the Doe case) and the International Labor Rights Fund (for the Roe case), provided legal counsel and support for the suits. Unocal responded by denying that the abuses occurred and insisting that, even if they had, the oil companies bore no responsibility for them. Company spokespeople stressed that Unocal's subsidiary was only an investor in the project and had no control over its daily operation.

Legal Issues

The basic issue in both suits was whether, and to what degree and under what circumstances, a company that invests in a project in another country is responsible for alleged violations of human rights committed by the country's government in connection with the project. They also highlighted a brief, obscure law called the Alien Tort Claims Act (ATCA), passed in 1789 to allow federal courts to hear small numbers of cases involving violations of international law such as piracy. This law (28 U.S.C. 1350) consists of only one sentence: "The district courts shall have original jurisdiction of any civil action by an alien for a tort only, committed in violation of the law of nations or a treaty of the United States."

Beginning with a 1980 case called *Filartiga v. Peña-Irala,* human rights organizations have successfully used ATCA to prosecute representatives of overseas governments for torture, genocide, and other abuses held to be egregious enough to violate international law. In 1991, furthermore, Congress passed the Torture Victim Protection Act, which specifically authorized this use of ATCA in cases alleging torture and other major rights violations by foreign governments. The outcome of *Doe v. Unocal* and related suits potentially could affect, or be affected by, the outcome of other ATCA suits.

Decision

As of mid-2004, the suits, or their descendants, are still in litigation. However, a number of important decisions have already been rendered during their progress through the courts.

Unocal moved for summary judgment in both suits—essentially, a dismissal—and a district court granted the company's request in September 2000, stating that although evidence suggested that Unocal knew about and benefitted from the military's use of forced labor on the project, the plaintiffs had not convincingly shown that the company had "participated in or influenced" the military's actions or "sought to employ forced or slave labor." However, the plaintiffs appealed the ruling. They also filed a suit in a California state court that was similar to the federal ones. Unocal moved to dismiss the state suit but was ultimately rejected in August 2001.

Los Angeles Superior Court judge Victoria Chaney made some of the most important rulings in the state suit (in which *Doe* and *Roe* were combined under the *Doe* name in June 2002). On the one hand, she awarded Unocal summary judgment on the charge of being directly liable for any human rights violations that might have occurred, saying that the company did not cause the alleged injuries either through intention or negligence. "In fact," she wrote, "the Joint Venturers expressed concern that the Myanmar government was utilizing forced labor in connection with the project." On the other hand, Chaney denied the oil company summary judgment on the charge of being indirectly, or vicariously, liable for the Myanmar's military actions, leaving this issue to be settled at trial. She also denied summary judgment on possible violation of several California laws.

In November 2003, Chaney granted Unocal's request to divide the state trial into phases, the first of which was to deal solely with the question of whether the MGTC, the Burmese company in charge of the project, was

Unocal's "alter ego," a necessary finding if the parent company rather than the subsidiary was to be sued or if "piercing the corporate veil" to reveal the exact relationship between parent and subsidiary was to be ordered. Chaney ruled on April 14, 2004, that this was not the case. She reiterated that Unocal could not be held directly responsible for the plaintiffs' alleged injuries, but she also found that "Unocal knew or should have known that there were human rights violations in Myanmar (Burma) at the time it entered into negotiations for the Yadana Project." On September 14, 2004, Chaney ruled that the second phase of the state trial, focusing on vicarious liability, could proceed.

Meanwhile, the Ninth Circuit Court of Appeals reviewed the 2000 decision to dismiss the two federal suits. A three-judge panel of the appeals court heard the case in September 2002 and partly reversed the lower court's decision. Saying that the plaintiffs needed only to show that Unocal knowingly "aided and abetted" the military in perpetrating the abuses, not that it controlled the military or directly caused or requested the abuses—a more lenient standard than the district court had applied—the panel ruled that the plaintiffs had presented enough evidence to justify a trial and the case should therefore return to the lower court. Unocal demanded a rehearing by the entire appeals court. In a surprise move in late 2004, just as the rehearing was scheduled to take place, both the state and the federal suits were settled out of court on terms that were not released.

Impact

Both sides in this dispute claimed victories among the decisions rendered in these cases. Unocal pointed to the success of several of its requests for summary judgment, especially the one finding it not guilty of intentional tort and negligence. The human rights groups, on the other hand, were heartened by Judge Chaney's statement that Unocal knew or should have known about the alleged abuses.

Because the suits were eventually settled out of court, however, they will provide no definitive ruling on whether oil companies can be held liable for governmental human rights abuses on their projects or whether ATCA can be used in suits against them.

AGUINDA ET AL. V. TEXACO, 142 F. SUPP. 2D 534 (SDNY 2001)

Background

Texas Petroleum (TexPet), a fourth-level subsidiary of the U.S.-based multinational oil company ChevronTexaco, owned 37.5 percent of the shares in a consortium headed by PetroEcuador, the Ecuadoran national oil company, which extracted oil from the country's Oriente region, a rain forest area in the Amazon River basin. TexPet was actively involved in the drilling from 1964 to 1990, during which time it allegedly discharged 18.5 billion gallons of "produced water"—water extracted with oil and left over when the process is complete—into rivers and streams in the area or left it in waste pits, from which it seeped into groundwater. In 1992, TexPet turned over its share of the operation to PetroEcuador, which became the project's sole owner and manager.

Natives of the Oriente, aided by environmental and human rights groups both within and outside the country, claim that this discharged water contained chemicals that poisoned people who used the river water for drinking, cooking, and other purposes, causing notable increases in skin rashes, spontaneous abortions (miscarriages), cancers, and other health problems. The water, they say, should have been reinjected into the wells, as is the practice in most countries. They also maintain that the Trans-Ecuadoran Pipeline, part of the same project, leaked 1.8 million gallons of crude oil, almost twice as much as spilled from the tanker *Exxon Valdez* during the famous Alaska disaster in 1989. They say that the project caused so much ill health and disruption that it amounted to "cultural genocide," resulting in the virtual extinction of the Cofan, Secoya, and Siona peoples.

ChevronTexaco denies these charges, saying that the health studies on which the groups' accusations are based do not constitute "credible, substantiated scientific evidence."[3] It points to the $40 million it spent to clean up the area between 1995 and 1998 and the fact that, at the completion of the cleanup, the government of Ecuador declared itself satisfied and released the company from all further liability. Furthermore, it says, audits by two "internationally recognized consulting firms" absolved its Amazon drilling operations of having caused any "lasting or significant environmental impact."[4]

The protesting groups filed suit in New York against Texaco, then headquartered in that city, on behalf of some 30,000 Ecuadorans in November 1993. This suit became known as *Aguinda v. Texaco* after Maria Aguinda, the first in the long list of Ecuadoran plaintiffs. A group of 25,000 Peruvian natives, living downstream on the Amazon and claiming similar health problems from the project's water pollution, filed a second suit, *Jota v. Texaco*, in 1994. The suits sought money damages on a variety of charges as well as a cleanup program to mitigate the alleged environmental damage. Ultimately the two suits were combined under the *Aguinda* name.

Legal Issues

The chief legal issue so far has been the question of where any trial resulting from the suits would be held. A legal doctrine called *forum non conveniens* (inconvenient venue) says that a trial must be held in the location most convenient to all parties. The plaintiffs tried to keep the trial in the United States, saying that the decisions that resulted in the alleged environmental damage were made in Texaco's New York headquarters. They also claimed that the judicial system in Ecuador, especially in rural areas like the one at issue, would not be able to handle such a complex case and that Ecuador's courts have usually taken the side of oil companies in disputes because so much of the country's revenue comes from petroleum. Texaco, on the other hand, held that Ecuador was the most convenient venue because having the trial there would allow easy interviews with witnesses and inspection of the supposedly damaged sites.

A second issue concerns the Alien Tort Claims Act (ATCA), a brief and enigmatic 1789 law invoked in some suits alleging involvement of U.S.-based oil companies in human rights abuses, such as *Doe v. Unocal.* The plaintiffs in *Aguinda* claimed that Texaco should be subject to this law because the company's actions violated international laws against environmental destruction, racial and ethnic discrimination (because the company used "primitive" waste disposal techniques in Ecuador that it would not have used in the United States), and cultural genocide. The Texaco suit raises the question of whether ATCA can be applied to redress alleged environmental and health damage as well as more obvious human rights abuses.

Decision

Texaco moved for dismissal of the *Aguinda* suit in December 1993 on the grounds of *forum non conveniens*, comity (defined in an 1895 Supreme Court decision as "the recognition which one nation allows within its territory to the legislative, executive or judicial acts of another nation," which would have given precedence to Ecuadoran law in this situation), and failure to join (include in the suit) indispensable parties (the Ecuadoran government and PetroEcuador).[5] The company's attorneys presented a letter from the Ecuadoran ambassador, saying that Ecuador considered the suit an affront to its national sovereignty. Nonetheless, district court judge Vincent Broderick ruled that the dismissal request was premature and told the parties to proceed with discovery to find out whether key decisions concerning the project were made in New York.

Judge Broderick died in March 1995, and the two cases were transferred to Judge Jed S. Rakoff, who dismissed them in 1996 and 1997 on the grounds Texaco had mentioned earlier. He was not persuaded by a motion to intervene filed by Ecuador, which had now changed its mind and was willing to have the trial take place in the United States, because it did not waive sovereign immunity. The plaintiffs appealed the dismissal, and in

1998 a three-judge panel of the Second Circuit Court of Appeals reversed it in the *Jota* case, sending the case back to Rakoff. The panel said that dismissal on the grounds of *forum non conveniens* and comity was in error because Texaco had not been required to submit to jurisdiction in the potential alternative forum, Ecuador, and the *forum non conveniens* doctrine could not be applied unless two possible sites were available for a trial. During this hearing, Texaco said it would accept an Ecuadoran trial.

In May 2001, Rakoff again dismissed the suits on the grounds of *forum non conveniens*, saying that the plaintiffs should refile them in Ecuador or Peru if they wished to continue them. He claimed that ATCA was unlikely to be applicable "because environmental torts are unlikely to be found to violate the law of nations" and that, in any case, the ATCA claim did not preclude the cases being tried in Ecuador. The plaintiffs appealed the decision again, and this time, on August 16, 2002, the appeals court affirmed Rakoff's ruling, saying that Ecuadoran courts were a quite acceptable venue and giving arguments against each of the plaintiffs' criticisms of them.

The suit against ChevronTexaco was refiled in Ecuador in May 2003 and began in the town of Lago Agrio in October. The plaintiffs are asking for $1 billion in damages, including an improved cleanup of the area and monitoring of long-term health effects of the alleged pollution. In late 2004, the trial was still proceeding.

Impact

The impact of the case will depend on what the Ecuadoran court decides. Its decision will affect the view of both rights groups and oil companies about the value of holding trials of this kind in the countries where the alleged abuses occurred, as opposed to in the United States. The Ecuadoran government's traditional support of the oil industry may be offset by considerable popular support for the plaintiffs. The verdict, especially if it goes against Texaco, could have a considerable impact on the way international oil companies invest and conduct projects in developing countries.

The court's negative comments about the relevance of ATCA reduces the likelihood that this controversial statute can be applied to environmental and health cases. However, it does not completely foreclose this possibility.

CHENEY V. U.S. DISTRICT COURT, 03-475 (2004)

Background

A few days after he took office in 2001, President George W. Bush ordered his vice president, Dick Cheney, to form a task force to design a national energy policy. During its deliberations, the task force, called the National Energy Policy Development Group, consulted with a number of private individuals, including some connected with oil and other energy industries. The task force published its recommendations five months after it was formed, then disbanded.

Environmental groups and other critics of the committee's policy proposals claimed that its advisers had influenced it to favor the energy industries unduly. They demanded to know the names of these advisers and see records of the committee's deliberations. "Once Congress and the American people get the details about what happened at the task force's closed-door meetings, the administration's energy plan will be revealed for what it is—a payback to corporate polluters," Sharon Buccino, an attorney for the Natural Resources Defense Council, said later.[6] Bush and Cheney refused to provide this information, citing executive privilege.

Later in 2001, the environmental group Sierra Club and a conservative watchdog organization called Judicial Watch filed a suit to obtain the information, citing the Federal Advisory Committee Act (FACA), a 1972 law that requires committees advising the president or vice president to meet in public and make their records accessible. The General Accounting Office (GAO), the research arm of Congress, filed a second suit, making the same demand—the first time this federal agency had done such a thing. The Natural Resources Defence Council (NRDC) filed a third suit, requesting the names of the advisers under the 1974 Freedom of Information Act.

Legal Issues

Basically, the issue was the degree to which executive privilege and the separation of powers specified in the Constitution protect the president and vice president and their advisers from public scrutiny and the circumstances, if any, under which the judiciary may order that protection breached. There was also the question of how higher courts should limit or dissolve the orders of lower ones or command actions from government officials. One possibility considered at several levels of the proceeding was the issuing of a so-called writ of mandamus ("we command"), which can be sent to a person, corporation, or inferior court within a higher court's jurisdiction and require a particular action to be taken. Courts normally do not use such a writ if other means of compelling action are available.

Interpretation of the Federal Advisory Committee Act was in question as well. Government attorneys claimed that the act was not applicable because it exempts committees "composed wholly of full-time, or permanent part-time, officers or employees of the federal government." The suing groups agreed that the Cheney task force itself fit this description, but they insisted that the law could still be used because "non-federal employees . . . regularly attended and fully participated in non-public meetings" and therefore could be considered "de facto members" of the committee. The Bush administration claimed that Congress did not intend for outside members to be considered and that, in any case, the act did not authorize suits against the president or vice president.

Finally, the case brought up the possibility of undue influence on formation of the country's potential energy policy. It weighed the president's and vice president's right to perform their duties unobstructed against the public's right to know the identity and workings of a key committee and its advisers.

Decision

In 2002, a district court dismissed some aspects of the Sierra Club–Judicial Watch suit but ruled that the plaintiffs could proceed with "tightly reined" pretrial discovery to determine whether any private citizens advising the energy committee could be considered to be de facto members of the group. The GAO suit was dismissed, and the agency did not refile it.

The Bush administration asked an appeals court to issue a writ of mandamus against the district court, requiring it to stop the discovery process and dismiss the vice president from the suit. In 2003, a panel of judges from the appeals court agreed that the range of information demanded by the suing groups was overly broad, going well beyond what FACA authorized, but it nonetheless denied the government's request, stating that it had no authority to apply the "extraordinary remedy of mandamus." The vice president could probably protect himself, it said, simply by claiming executive privilege in the district court. It stated that the separation of powers did not necessarily protect him, however, citing the 1970s case in which a court had forced President Richard Nixon to give up the Watergate tapes. The government asked the Supreme Court to review the case, and the Court agreed to do so in December 2003. It heard arguments on April 27, 2004, and delivered its decision on June 24. By a vote of 7 to 2, the high court reversed the appeals court's ruling.

Justice Anthony Kennedy wrote the Court's majority opinion. He concluded that the appeals court did have the authority to issue a writ of mandamus because of the importance of protecting the executive branch from "vexatious litigation" and preserving the confidentiality of its actions and deliberations. The separation of powers doctrine, he said, "affords presidential confidentiality the greatest protection consistent with the fair administration of justice" and does not depend upon executive privilege being invoked. Kennedy claimed that Nixon's situation had been different from Cheney's because the Nixon case was part of a criminal investigation. He also criticized the anti-Cheney groups for "ask[ing] for everything under the sky" and thus placing an undue burden on the executive branch, whereas the requests in the Nixon case had been much more narrow and specific.

Kennedy did not order the appeals court to issue a mandamus writ against the district court, as the government had requested. Instead, he returned the case to that court for further consideration of the government's mandamus petition, advising the judges to pay special attention to the "weighty separation-of-powers objections" that the administration had expressed.

Justices Ruth Ginzberg and David Souter filed a dissenting opinion. Justice Ginzberg, who wrote the opinion, said she would have affirmed the appeals court's decision to deny the mandamus writ because the government was not asking for a narrowing of the organizations' discovery requests, which she thought would have been justified, but rather for a blocking of all discovery, which she did not support. She believed that the "tightly reined in" discovery authorized by the district court should have been allowed to proceed, with the understanding that the government could make objections to releasing particular pieces of information if it wished to do so.

Impact

The Supreme Court's decision protects Vice President Cheney and other officials from releasing the contested documents at present and makes it unlikely that they will ever have to do so. However, it does not completely rule out that possibility. The lower courts could still permit the discovery process, or the information might be obtained without directly involving the vice president. On April 1, 2004, for instance, ruling in the NRDC suit, district court judge Paul Friedman ordered the Energy and Interior Departments to release thousands of pages of documents related to the energy task force's deliberations. He said that these departments were not shielded by executive privilege.

SOSA V. ALVAREZ-MACHAIN ET AL. 03-339 (2004)

Background

In 1985, a Drug Enforcement Administration (DEA) agent named Enrique Camarena-Salazar was captured during an assignment in Mexico, tortured, and killed. Humberto Alvarez-Machain, a Mexican physician, was alleged to have helped to extend Camarena's life so he could be tortured longer. After a California grand jury indictment, a district court issued a warrant for Alvarez's arrest, but Mexico refused to extradite him. The DEA then hired a group of Mexicans led by José Francisco Sosa to kidnap Alvarez and bring him to the United States for trial. In 1990, Sosa's group captured Alvarez, held him overnight at a motel, and then flew him to El Paso, Texas, where he was arrested.

When brought before a judge, Alvarez held that he should be released because his kidnapping was "outrageous governmental conduct" and violated the extradition treaty between Mexico and the United States. The district court and the Ninth Circuit Court of Appeals agreed, but the Supreme Court reversed their decisions. Alvarez was ultimately put on trial in 1992 and acquitted. He returned to Mexico and from there filed suit in 1993 against Sosa and the others who had kidnapped him, several DEA agents, and the United States.

Legal Issues

Alvarez sued for false arrest under the Federal Tort Claims Act (FTCA), which authorizes suit "for . . . personal injury . . . caused by the negligent or wrongful act or omission of any employee of the Government while acting within the scope of his office or employment," and under the Alien Tort Claims Act (ATCA), called in the court papers the Alien Tort Statute (ATS), the brief law passed in 1789 that gives federal district courts jurisdiction over suits brought by aliens (noncitizens) for tort (injury) actions "committed in violation of the law of nations." Because the courts' decisions in the Alvarez case affect the general applicability of ATCA to cases involving alleged human rights violations overseas, they have profound implications for suits such as *Doe v. Unocal*, in which rights groups are attempting to use ATCA against oil companies that they accuse of abetting such violations during projects in developing countries.

Decision

A district court granted the government's motion to dismiss the FTCA claim but awarded a summmary judgment and $25,000 damages to Alvarez for violation of the ATS (ATCA). A three-judge panel of the Ninth Circuit appeals court upheld the ATS judgment and also reopened the possibility of an FTCA case. At the government's request, the entire appeals court reheard the case. Their vote was divided, but the majority reached the same decision that the panel had. In regard to the ATS it cited the "clear and universally recognized norm prohibiting arbitrary arrest and detention" as a reason for concluding that Alvarez's arrest could be considered a tort in violation of international law.

The government appealed the case to the U.S. Supreme Court, which agreed to hear it in order to "clarify the scope of both the FTCA and the ATS." The high court gave its decision on June 29, 2004, ruling that Alvarez was "not entitled to a remedy under either statute."

Justice David Souter delivered the majority opinion of the court. He discussed extensively the history of the ATS/ATCA, about which he said the opposing parties in the suit had "radically different . . . interpretations." Alvarez and his counsel maintained that "the ATS was intended not simply as a jurisdictional grant, but as authority for the creation of a new cause of action for torts in violation of international law." Souter dubbed this conclusion "implausible" and wrote instead that the law was more likely to have been intended only to grant jurisdiction in a tiny number of unusual cases. (He noted that between the time of the law's passage and 1980 it was successfully called upon only once.) On the other hand, he also rejected Sosa's (and the Bush administration's) position that the law was essentially meaningless.

Souter concluded that the only crimes that would have been recognized as qualifying for suits brought under this brief law when it was passed as part of the Judiciary Act of 1789 were piracy on the high seas, infringement of the rights of foreign ambassadors, and violation of safe conducts. All were violations of international law "admitting of a judicial remedy and at the same time threatening serious consequences in international affairs." He noted that "there is no record of congressional discussion about private actions that might be subject to [this] jurisdictional provision."

Souter did not completely reject the possibility of private suits under ATCA, but he held that "courts should require any claim based on the present-day law of nations to rest on a norm of international character accepted by the civilized world and defined with a specificity comparable to the features of the 18th-century paradigms we have recognized." He did not say what crimes, if any, might qualify. However, he stated that Alvarez's claim of "arbitrary arrest" and forcible detention did not meet this requirement because the only forcible detention Alvarez suffered before being in a jurisdiction that had a legal right to hold him was his overnight stay at the Mexican motel.

Continuing with his general discussion of ATCA suits, Souter pointed out that "this Court has recently and repeatedly said that a decision to create a private right of action is one better left to legislative judgment" than to the courts "in a great majority of cases." For one thing, he said, allowing such suits could have "potential implications for the foreign relations of the United States," which are under the control of the legislative and executive branches of government. He cited a 1984 opinion by Justice Robert H. Bork expressing doubt that ATCA should be interpreted to require "our courts [to] sit in judgment of the conduct of foreign officials in their own countries with respect to their own citizens."

Souter did note that the Torture Victim Protection Act, passed in 1991, "establish[es] an unambiguous and modern basis for" suits alleging torture and extrajudicial killing in other countries. This law has been considered a sort of supplement to or expansion of ATCA, and its legislative history included the remark that ATCA should "remain intact to permit suits based on other norms that already exist or may ripen in the future into rules of customary international law." To be sure, Souter pointed out, Congress "has done nothing to promote such suits," and indeed, the Senate several times had "expressly declined to give the federal courts the task of interpreting and applying international human rights law." Still, he maintained that "the door" of ATCA suits "is still ajar subject to vigilant doorkeeping, and thus open to a narrow class of international norms today." He added, "It would take some explaining to say . . . that federal courts must avert their gaze entirely from any international norm intended to protect individuals."

Impact

Souter's ruling rejected the applicability of ATCA to Alvarez's case, and his general discussion certainly did not encourage private ATCA suits such as *Doe v. Unocal.* Nonetheless, it heartened some groups trying to bring such suits because it did not foreclose them entirely. It left up to judges the decision of which "causes of action" might qualify for suits under this law—a notion that Justice Antonin Scalia, seconded by the chief justice, William H. Rehnquist, and Justice Clarence Thomas, strongly objected to in a concurring opinion, stating that "the judicial lawmaking role [this part of the ruling] invites would commit the federal judiciary to a task it is neither authorized nor suited to perform."

"The *Alvarez* decision erases any doubt about the validity of the ATCA for addressing egregious human rights cases, and sends a clear message to multinationals who seek to profit from forced labor and torture of workers and other human rights victims," Terry Collingsworth, executive director of the International Labor Rights Fund, claimed after the decision.[7] Conversely, Daniel Petrocelli, an attorney with the firm of O'Melveny & Meyers in Los Angeles, said that the decision was "a nail in the coffin of [ATCA] being used against U.S. businesses."[8] In fact, the ruling did not cover the issue of corporate liability under ATCA at all. Souter referred to such a possibility only once, in case footnote number 20, in which he mentioned the question of "whether international law extends the scope of liability for a violation of a given norm to the perpetrator being sued, if the defendant is a private actor such as a corporation or an individual." He offered no answer to this question, however, so rulings on it will have to wait for future cases.

[1] Energy Information Administration, "Public Utility Holding Company Act of 1935: 1935–1992." Washington, D.C.: Energy Information Administration, 1993, p. 1.

[2] Fred Bosselman, Jim Rossi, and Jacqueline Lang Weaver. *Energy, Economics and the Environment: Cases and Materials.* New York: Foundation Press, 2000, p. 718.

[3] "Texaco in Ecuador: Response to Claims." Texaco web site. Available online. URL: http://www.texaco.com/sitelets/ecuador/en/response_to_claims/. Accessed on July 10, 2004.

[4] "Texaco in Ecuador: Background." Texaco web site. Available online. URL: http://www.texaco.com/sitelets/ecuador/en/overview/. Accessed on July 10, 2004.

[5] Bosselman, Rossi, and Weaver, *Energy, Economics and the Environment*, p. 1,170.

[6] Sharon Buccino, quoted in David G. Savage, "Judge Gives Deadline on Energy Task Force Records," *Los Angeles Times*, reprinted in *San Francisco Chronicle*, April 2, 2004, p. A7.

[7] Terry Collingsworth, quoted in William Baue, "Does Supreme Court Validation of Alien Tort Claims Act Apply to Corporations?" CSRWire (Corporate Social Responsibility Newswire Service). Available online. URL: http://www.csrwire.com/sfarticle.cgi?id=1458. Posted on July 1, 2004.

[8] Daniel Petrocelli, quoted in Baue, "Does Supreme Court Validation of Alien Tort Claims Act Apply to Corporations?"

CHAPTER 3

CHRONOLOGY

440–45 million years ago

- Period during which most scientists believe fossil fuels (coal, oil, and natural gas) were formed.

circa 600 B.C.

- The Chinese drill the first oil wells and make the first oil pipelines.

1765

- Scottish inventor James Watt creates an engine that uses steam at high pressure to move a piston up and down in a cylinder. This engine proves capable of powering many kinds of machines and helps to make the Industrial Revolution possible.

1789

- Congress passes the Alien Tort Claims Act or Alien Tort Statute, a law allowing federal courts to try a small number of cases involving alleged violations of international law. Some 200 years later, this law will be applied to cases involving torture, genocide, and other overseas human rights violations, including some allegedly abetted by American oil companies.

1831

- British scientist William Faraday discovers that by moving a magnet around a wire (or a wire within a magnetic field), alternating electric current can be made to flow in the wire. This is the principle behind most modern ways of generating electricity.

1839

- British physicist William Grove invents the hydrogen fuel cell.

Chronology

1857

- American inventor Michael Dietz creates a clean-burning, odorless lamp that uses kerosene as a fuel; it quickly becomes very popular, causing kerosene to replace whale oil as the standard lighting fuel.

1859

- ***August 27:*** "Colonel" Edwin Drake strikes oil in his well near Titusville, Pennsylvania, the first oil well in the United States.

1870

- John D. Rockefeller and others form the Standard Oil Company of Ohio, which establishes a monopoly by controlling all aspects of oil refining and distribution.

1874

- French science fiction writer Jules Verne publishes a book called *The Mysterious Island,* which describes an economy based on the use of water as a fuel. This idea is similar to that of the "hydrogen economy" later proposed by Jeremy Rifkin and others, in which the hydrogen is made from water with energy from renewable sources.

1876

- The Supreme Court rules in *Munn v. Illinois* that a state government has the right to regulate grain warehouses and other private businesses "clothed with a public interest," including setting maximum prices that the businesses can charge. This ruling is later applied to gas and electric utilities.

1879

- Thomas A. Edison invents the electric light.

1880

- The first U.S. hydroelectric power plant is set up in Wisconsin.

1882

- Thomas Edison opens the first electricity generating stations in London and New York.
- John D. Rockefeller and his partners form the Standard Oil Trust to control the numerous companies they own.

1890

- Congress passes the Sherman Antitrust Act to prevent monopoly control by trusts such as John D. Rockefeller's Standard Oil Trust, which controls the oil industry.

1892

- U.S. government regulators force the Standard Oil Trust to break up.

1899

- John D. Rockefeller transfers the assets of the former Standard Oil Trust, including controlling blocks of stock in supposedly separate companies, to the Standard Oil Company of New Jersey.

1901

- ***January 10:*** Spindletop, an oil field near the town of Beaumont in southeastern Texas, opens with a "gusher" that shoots oil more than 150 feet into the air. The field is soon producing almost 100,000 barrels a day.

1906

- The Justice Department charges Standard Oil of New Jersey and related defendants with violating the Sherman Antitrust Act.

1911

- ***May 15:*** The U.S. Supreme Court upholds a lower court's decision to break up the holdings of the Standard Oil Company of New Jersey because the company has violated the Sherman Antitrust Act.

1912

- Winston Churchill, First Lord of the Admiralty, decrees that British navy ships will hereafter use oil rather than coal as their fuel.

1913

- The world's first geothermal electric power plant is commissioned in Larderello, Italy.

1914–1918

- Use of oil rather than coal gives British navy ships an advantage during World War I.

Chronology

1920

- Congress passes the first federal energy laws, the Federal Water Power Act (which establishes the Federal Power Commission to set up and oversee government hydroelectric projects), and the Mineral Leasing Act (which sets up rules for leasing federal lands for oil and gas extraction).

1930

- Iceland begins the first large-scale use of geothermal power for space heating.

1935

- Congress passes the Public Utility Act. It includes the Public Utility Holding Company Act (PUHCA), which regulates ownership and stock holdings of public utilities and orders private electric utility holding companies to be broken up by 1955, and the Federal Power Act, which gives the Federal Power Commission the right to regulate wholesale electric utility rates.

1938

- Congress passes the Natural Gas Act, which gives the Federal Power Commission control of rates for interstate pipeline transmission of natural gas.
- Atomic fission, the process later harnessed in nuclear power plants, is demonstrated in Germany.

1939–1945

- Oil shortages among the Axis powers and ample supplies among the Allies influence the outcome of World War II.

1954

- Bell Laboratories invents the modern solar (photovoltaic) cell.
- In a key case, *Phillips Petroleum v. Wisconsin*, the U.S. Supreme Court rules that the Federal Power Commission has the right to regulate wellhead prices for natural gas shipped between states as well as pipeline transmission prices.
- ***September 16:*** In a speech to a group of science writers in New York, Atomic Energy Commission chairman Lewis Strauss claims that in a generation or so, electricity produced in nuclear plants will be "too cheap to meter."

Energy Supply

1957

- The first nuclear power plant in the United States begins operation at Shippington, Pennsylvania.

1959

- British engineer Francis Thomas Bacon invents the modern hydrogen fuel cell.

1960

- Saudi Arabia, Iran, Iraq, Kuwait, and Venezuela form the Organization of Petroleum-Exporting Countries (OPEC).

1969

- For the first time, the United States imports more crude oil than it exports.
- An accident at an offshore oil rig near Santa Barbara, California, releases a gigantic oil slick that fouls the shore and harms wildlife. Outrage about this spill leads to legislation that places much of the California coast off-limits to drilling.

1970

- Production of petroleum in the continental United States (excluding Alaska) peaks at 9.4 million barrels per day, bearing out a prediction by geophysicist M. King Hubbert.
- Congress passes the version of the Clean Air Act that basically still applies.

1973

- ***February 22:*** The U.S. Supreme Court rules in *Otter Tail Power Co. v. United States* that utilities are subject to antitrust laws and that courts can order them to "wheel" power from generators that they do not own.
- ***October 6:*** Egypt and Syria, with the aid of other Arab countries, attack Israel on the Jewish holy day of Yom Kippur. A brief war erupts, during which Israel takes back the territory that was seized from it.
- ***October 17:*** As punishment for American support of Israel during the "Yom Kippur War," Arab countries stop shipping oil to the United States.

1974

- The United States and 15 other countries form the International Energy Agency as a counterbalance to OPEC.
- ***March:*** The Arab oil embargo against the United States ends.

Chronology

1975

- The United States begins phasing out the use of lead as a gasoline additive.
- ***December 22:*** The Energy Policy and Conservation Act becomes law. Among other measures, it sets up the Corporate Average Fuel Economy (CAFE) program, which establishes separate minimum fuel efficiency (in miles per gallon) standards for passenger cars and light trucks, to be met over the average of a vehicle manufacturer's whole line. It also establishes the Strategic Petroleum Reserve, an emergency supply of hundreds of millions of barrels of oil.

1977

- Congress passes the Foreign Corrupt Practices Act, which forbids U.S. citizens or corporations to pay bribes to foreign officials to obtain or retain business, something that oil companies are often accused of doing.
- ***July 13–14:*** A widespread electrical blackout in New York City is accompanied by looting, rioting, and arson, resulting in much property damage and a number of injuries and deaths.
- ***August:*** President Jimmy Carter signs the DOE Organization Act, creating the Department of Energy (DOE). The Federal Energy Regulatory Commission (FERC), which replaces the Federal Power Commission as the federal agency that regulates energy, is established as part of the DOE.

1978

- Congress passes the Public Utility Regulatory Policies Act (PURPA), which requires electric utilities to buy power offered by "qualifying facilities" (QFs)—cogenerators and small nonutility generators that use renewable resources. Although intended primarily to encourage conservation and the use of renewable energy sources, this act also proves to be the first step toward deregulating and restructuring the electricity industry to make it more competitive.
- Congress passes the Natural Gas Policy Act, which begins deregulation of the natural gas industry by reducing government control of wellhead prices.

1979

- The oil tankers *Atlantic Empress* and *Aegean Captain* collide in the Caribbean, spilling 110 million gallons of oil.
- Brazil requires all cars and other light vehicles in the country to use either ethanol or gasohol (a combination of ethanol and gasoline) as a fuel.

- A rebellion ousts the shah of Iran, who has ties to the United States, and replaces his government with an anti-American, Islamic fundamentalist one.
- ***March 28:*** A combination of mechanical failure and human error at a nuclear power plant in Three Mile Island, Pennsylvania, produces an accident in which a small amount of radiation is leaked into the surrounding air and water. This radiation release causes no provable damage but greatly heightens public fear of nuclear power.
- ***June 3:*** A gigantic blowout begins at a Mexican exploratory offshore well, Ixtoc I, in the Bay of Campeche, part of the Gulf of Mexico. By the time the blowout is stopped almost 10 months later, it has put more than 140 million gallons of oil into the water. This is the largest single oil spill in history.
- ***November:*** After the new government of Iran seizes and holds several Americans as hostages, President Carter bans the importation of Iranian oil. Iran, in turn, levies an oil embargo against the United States. This leads to a second worldwide "oil shock," like the one that followed the 1973–74 embargo.

1980

- In *Filartiga v. Peña-Irala*, an appeals court upholds the use of the 1789 Alien Tort Claims Act to allow a U.S. federal court to hear cases involving alleged torture by governments overseas. This is later used as a precedent in suits against U.S. oil companies for abetting human rights violations in developing countries where the companies had projects.
- ***January 23:*** President Carter tells Congress that the United States is willing to use military force to protect the oil-rich Persian Gulf region from takeover by outside forces. This policy statement later becomes known as the Carter Doctrine.

1982

- Congress passes the Nuclear Waste Policy Act, which orders the Department of Energy to set up a program for disposal of radioactive waste from nuclear power plants, such as spent fuel rods. Owners of such power plants are required to contribute to a fund that will pay for the disposal.

1985

- Minimum fuel economy standards for passenger cars and light trucks, established by Congress in 1975 as part of the Corporate Average Fuel Economy (CAFE) program, go into effect.

- FERC issues Order 436, which begins separating natural gas production, transmission, and distribution from one another.
- A world oil glut has painful economic effects.

1986

- FERC issues Order 451, which continues the unbundling of the natural gas industry.
- ***April 25:*** A major accident at a nuclear power plant in Chernobyl, Ukraine, then part of the Soviet Union, releases large amounts of radiation into the atmosphere.

1987

- An amendment to the Nuclear Waste Policy Act of 1982 directs the secretary of energy to investigate Yucca Mountain, Nevada, as a site for permanent underground disposal of the nation's radioactive waste from nuclear power plants.

1989

- Congress passes the Natural Gas Wellhead Decontrol Act, which states that all federal price controls on natural gas will end in 1993.
- ***March 24:*** The oil tanker *Exxon Valdez* runs aground in Prince William Sound in Alaska, spilling 10.8 million gallons of oil and creating widely publicized environmental damage.

1990

- The Clean Air Act is revised to, among other things, establish a "cap-and-trade" acid rain prevention program. The revision also requires oxygen-containing substances to be added to gasoline sold in the winter in some cities in order to reduce carbon monoxide emissions.
- Great Britain begins opening its electricity market to increased competition.
- In response to the *Exxon Valdez* spill, Congress passes the Oil Pollution Act, which mandates many measures to prevent and clean up oil spills.
- ***August 2:*** Military forces from Iraq invade the small neighboring country of Kuwait.

1991

- Congress passes the Torture Victim Protection Act, which confirms that U.S. courts can try cases involving alleged torture committed by foreign governments, using the Alien Tort Claims Act.

- ***January 17:*** Backed by a UN Security Council resolution, the United States leads a coalition of troops from several nations in an invasion of Iraq that becomes known as Operation Desert Storm, or the Gulf War. They quickly force Iraq to leave Kuwait.

1992

- Congress passes the National Energy Policy Act (EPAct), which, among other things, expands the definition of independent power producers and strengthens FERC's authority to require owners of interstate electric transmission lines to carry ("wheel") electricity from third-party generators.
- Representatives of countries around the world meet at Rio de Janeiro, Brazil, at what is called the Earth Summit. Among other things, they agree to the Framework Convention on Climate Change, which calls for a reduction in greenhouse gas emissions worldwide in order to prevent global warming.
- ***April:*** The Federal Energy Regulatory Commission issues Order 636, which requires natural gas pipeline companies to sell gas, transportation services, and storage services separately ("unbundling") and transport gas owned by others, creating a fully competitive market for natural gas.

1993

- ***November:*** A suit, *Aguinda et al. v. Texaco*, is filed in New York on behalf of 30,000 Ecuadorans, claiming that a subsidiary of the oil giant caused considerable environmental damage during a drilling project in Ecuador.

1996

- Two groups file suit against Unocal, a California-based oil company, for alleged complicity in human rights abuses during construction of a gas pipeline in Burma (Myanmar) in which Unocal, through a subsidiary, was a minority investor.
- The European Union issues Directive 96/92/EC, which requires unbundling of electric generation, transmission, and distribution; open access to transmission facilities for all generators; and establishment of a transmission system operator. Full retail competition is to be in place by 2007.
- ***May 10:*** The Federal Energy Regulatory Commission issues Order 888, which requires utilities to transmit electricity from all generators at the same rate. It allows the utilities to recover stranded costs associated with following this mandate. It also recommends that regions of the country establish independent organizations to manage electricity transmission.

- ***September:*** Governor Pete Wilson signs legislation that makes California the first state to begin deregulating its wholesale electricity industry.

1997

- Member countries of the Organization for Economic Cooperation and Development sign the Convention on Combating the Bribery of Foreign Public Officials in International Business Transactions, making such bribery a criminal offense.
- ***December:*** Representatives of about 180 countries meet in Kyoto, Japan, to draft a plan for reduction of greenhouse gas emissions that becomes known as the Kyoto Protocol.

1998

- California's electricity deregulation rules go into effect.

1999

- Russia invades Chechnya after rebels from that country attack neighboring Dagestan. Michael Klare of Hampshire College in Amherst, Massachusetts, claims that one motive for Russia's action is to keep control of an area through which a pipeline from the oil-rich Caspian Sea area to the Black Sea might pass.
- U.S.-based multinational oil companies Exxon and Mobil merge to become ExxonMobil.

2000

- ***May:*** Increased demand and other factors lead to a sharp rise in the wholesale price of electricity in California's spot market.
- ***September:*** Britain launches the Renewables Obligation, a combined renewable portfolio standard and trading program that requires suppliers to purchase a certain percentage of their electricity from generators that use renewable sources, and the Climate Change Levy, a tax on industrial energy users and suppliers of energy products that will be used to help the country make the reduction in carbon dioxide emissions required by the Kyoto Protocol.

2001

- Representatives of 178 countries work out details of implementing the Kyoto Protocol at meetings in Bonn, Germany, in July and Marrakesh, Morocco, in November.

- George W. Bush appoints Vice President Dick Cheney to head a task force to develop a national energy policy. The task force makes its recommendations and disbands five months later.
- Several groups sue Cheney, demanding to know the names of private individuals who advised his task force and other details of its deliberations.
- U.S.-based multinational oil companies Chevron and Texaco merge to become ChevronTexaco.
- ***January:*** Rolling blackouts occur throughout California. Governor Gray Davis declares a state of emergency, and the state government halts deregulation. The California Department of Water Resources replaces the state's utilities as the buyer of wholesale electricity.
- ***February:*** Governor Davis begins signing approximately $44 billion worth of long-term contracts with energy suppliers, paying exceptionally high rates to guarantee the state's electricity supply.
- ***March 28:*** The European Union ratifies the Kyoto Protocol, but the U.S. government says that it will not do so.
- ***April:*** Pacific Gas and Electric, the investor-owned utility that provides power to Northern California, declares bankruptcy because of losses stemming from the California energy crisis.
- ***June 18:*** FERC temporarily caps wholesale electricity prices in 11 western states, including California.
- ***September:*** California's energy crisis is considered over.
- ***November:*** President Bush orders the Department of Energy to fill the Strategic Petroleum Reserve to full capacity.
- ***December 3:*** Energy marketing giant Enron Corporation declares bankruptcy.

2002

- A district court allows groups suing Vice President Cheney to proceed with "tightly reined in" discovery to determine whether private individuals who advised his energy task force can be considered de facto members of the committee.
- ***June:*** California judge Victoria Chaney rules in *Doe v. Unocal*, a state suit, that oil company Unocal is not directly liable for human rights abuses allegedly committed by the Burmese (Myanmar) military during construction of a gas pipeline in that country, but she also states that the company may be vicariously (indirectly) liable.
- ***July 9:*** Congress votes by a narrow margin to approve Yucca Mountain, Nevada, as the site for permanent disposal of radioactive waste from nuclear power plants.
- ***August 16:*** An appeals court upholds district court's rulings that the trial in *Aguinda v. Texaco* should be held in Ecuador.

- ***December:*** A strike beginning in Venezuela's state-owned oil company temporarily shuts down the country's oil production and exports, helps to cause an abrupt rise in oil prices, and almost topples the regime of President Hugo Chávez.

2003

- The Bush administration waives the Clean Air Act's "new source review" requirement for remodeled power plants.
- A panel of judges from an appeals court refuses the Bush administration's request to order a lower court to stop investigation in a suit attempting to force Vice President Cheney to reveal the identity of people who advised his energy policy task force.
- ***January:*** In his State of the Union address, President Bush promises $1.7 billion in funding for hydrogen fuel cell research and development.
- ***March 19:*** The United States begins an invasion of Iraq, an action widely claimed to be inspired at least partly by a desire to control that nation's considerable oil wealth. The government denies this claim.
- ***March 29:*** The Federal Energy Regulatory Commission releases papers proving that Reliant Resources, Inc., and dozens of other energy companies deliberately cheated California during the state's 2000–01 energy crisis. The commission orders the companies to refund $3.3 billion to the state.
- ***April:*** The National Highway Traffic Safety Administration agrees to revise the CAFE standard for light trucks upward by 1.5 mpg by 2007.
- ***April 6:*** Merchant banker James Giffen and former Mobil executive Bryan Williams are indicted for allegedly giving $78 million in bribes to the president of Kazakhstan (a country bordering the Caspian Sea) on behalf of several large oil companies, a violation of the Foreign Corrupt Practices Act.
- ***June:*** FERC rules that the state of California must honor the expensive long-term contracts it made with energy suppliers during its 2000–01 energy crisis, even though the contracts were partly the result of fraudulent market manipulation. The agency also revokes Enron Corporation's right to take part in electricity and natural gas markets.
- ***August 14:*** The worst blackout in U.S. history darkens New York, much of the Midwest, and Ontario, Canada, highlighting defects in the North American electricity transmission system.
- ***October:*** Trial in *Aguinda v. Texaco* begins in Ecuador.

2004

- ***January:*** Royal Dutch/Shell admits having overstated its proven oil reserves by 20 percent.

- ***March:*** More than 50 Representatives ask President Bush to stop the filling of the Strategic Petroleum Reserve temporarily so that more oil will remain in the market to buffer record high gasoline prices. The request is denied.
- ***April 5:*** An investigative panel appointed by the governments of the United States and Canada claims that much of the blame for the August 2003 northeastern blackout lies with FirstEnergy Corporation, a large utility in whose territory the transmission failure started.
- ***April 12:*** Pacific Gas and Electric, the Northern California utility that declared bankruptcy in the wake of the 2000–01 California energy crisis, emerges from bankruptcy after its financial restructuring plan is accepted. It (and its customers) must still pay off enormous debts during the next nine years.
- ***June 14:*** Snohomish County (Washington State) Public Utility District releases audiotapes, obtained in connection with a suit against Enron, that show Enron traders in August 2000 laughing about stealing money from "Grandma Millie" in California through fraudulent reports about electricity supplies and transmission.
- ***June 18:*** al-Qaeda militants in Saudi Arabia behead Paul Johnson, a U.S. contractor working in the country, whom they had previously kidnapped. This and several other attacks on foreigners in Saudi Arabia cause an increase in oil prices due to fears that the Saudi government can no longer protect employees of its trading partners.
- ***June 24:*** The Supreme Court rules by a vote of 7 to 2 that an appeals court should reconsider the government's request to deny watchdog groups the chance to obtain the names of people who met with Vice President Cheney's energy policy task force, saying that the separation of powers gives the president and vice president the right to keep the identity of their advisers private.
- ***June 28:*** U.S. forces return sovereignty of Iraq to an interim Iraqi government.
- ***June 29:*** A Supreme Court decision in *Sosa v. Alvarez-Machain* suggests that successful use of the Alien Tort Claims Act to prosecute U.S. citizens or corporations, including oil companies, for abetting human rights violations in other countries is unlikely but not impossible.
- ***July 9:*** A three-judge panel of a federal appeals court rejects the Nevada state government's constitutional challenge to the Yucca Mountain nuclear waste disposal project but orders the Environmental Protection Agency to devise an improved plan to protect the public against radiation releases beyond the 10,000-year time span currently considered.
- ***September 14:*** Los Angeles Superior Court judge Victoria Chaney rules that the second phase of the state trial in *Doe v. Unocal*, which will focus

on whether Unocal was "vicariously responsible" for the Burmese government's alleged human rights abuses on the site of a project in that country partly sponsored by the oil company, can go forward.

- ***November 4:*** Russia ratifies the Kyoto Protocol, the international treaty aimed at reducing emissions of carbon dioxide and other gases believed to contribute to global warming.
- ***late 2004:*** State and federal suits against Unocal, alleging complicity of the oil company's subsidiary in governmental human rights abuses during a gas pipeline project in Myanmar, are settled out of court by an undisclosed arrangement.
- ***early December:*** A 30-acre solar installation goes into operation near Muhlhausen, Germany, becoming the largest solar energy plant in the world.

2005

- ***January 30:*** Elections held in Iraq.
- ***February 16:*** Kyoto Protocol takes effect as international law.

CHAPTER 4

BIOGRAPHICAL LISTING

Francis Thomas Bacon, British engineer. In 1959, he invented the modern hydrogen fuel cell.

Colin J. Campbell, British geologist and petroleum industry consultant. Like Kenneth Deffeyes and some others, Campbell believes that the peak of world oil production will be reached before 2010 and may have already occurred.

Jimmy Carter, president of the United States, 1977–81. In response to two Arab oil embargoes, Carter introduced massive energy policy changes stressing conservation and diversification of energy sources. He also expressed willingness to protect the Middle East with military force if necessary, an idea that became known as the Carter Doctrine.

Dick Cheney, vice president of the United States, 2000–present. In 2001, soon after taking office, President George W. Bush appointed Cheney to form and lead a task force called the National Energy Policy Development Group. It made recommendations for a national energy policy later in the year and then disbanded. Several environmental and watchdog groups accused the committee of having been unduly influenced by outside consultants representing energy industry interests and sued Cheney, a former CEO of energy contractor Halliburton Company, to obtain the names of the people the task force had consulted. Cheney refused, citing executive privilege, and a June 2004 Supreme Court decision upheld his right to do so.

Winston Churchill, British political leader. In 1912, as First Lord of the Admiralty, he ordered British navy ships to use oil instead of coal. This change gave the British an advantage during World War I and greatly increased the popularity of oil as a fuel.

Gray Davis, governor of California, 1999–2003. Davis's actions during the California energy crisis of 2000–01, which included signing some $44 billion worth of long-term contracts with electricity suppliers in February 2001, were criticized by some and probably played a role in his recall in October 2003.

Biographical Listing

Kenneth S. Deffeyes, petroleum geologist and emeritus professor at Princeton University. A former colleague of famed geophysicist M. King Hubbert, Deffeyes has applied Hubbert's calculations to world oil production and concluded that production will peak in the first decade of the 21st century.

Michael Dietz, U.S. inventor. In 1857, he created a clean-burning, odorless lamp that used kerosene as a fuel. This lamp became very popular, making kerosene and therefore, indirectly, oil itself valuable.

"Colonel" Edwin Drake, U.S. entrepreneur. On August 27, 1859, he struck oil in a well near Titusville, Pennsylvania, the first oil well in the United States.

Thomas A. Edison, prolific U.S. inventor. He created the electric light in 1879 and established the first electricity generating plants in London and New York in 1882.

Michael Faraday, British scientist. In 1831, he showed that when a magnet is moved around a wire or a wire turns in a magnetic field, an electric current flows in the wire. This observation made bulk electricity generation possible.

James Giffen, CEO of Mercator Corporation, a New York merchant bank. Giffen was indicted in April 2003 for allegedly giving $78 million in bribes to the president of Kazakhstan, a Caspian Sea country, on behalf of several oil companies in violation of the Foreign Corrupt Practices Act. He has declared his innocence.

William Grove, British physicist. In 1839, he invented the first hydrogen fuel cell, which he called a gas battery.

M. King Hubbert, celebrated U.S. geophysicist. Hubbert concluded that oil production follows a bell curve. Using calculations based on this idea, he claimed in 1956 that oil production in the United States would peak in the early 1970s and decline permanently thereafter. His ideas were ridiculed at the time, but they attracted widespread attention after it became clear that production in fact did peak in 1970.

Saddam Hussein, president of Iraq, 1979–2003. In 1990, he invaded neighboring Kuwait, possibly with the aim of controlling its oil wealth. A UN coalition led by the United States drove him out a few months later. The United States invaded Iraq beginning in March 2003, claiming that Hussein possessed weapons of mass destruction, and captured him on December 13, 2003.

Samuel Insull, president of Commonwealth Edison Company. A former assistant to Thomas Edison, Insull helped to create a system for transmitting electricity over long distances from central power stations. He also developed the concept of the electric utility as a vertically integrated monopoly.

Mikhael Khodorkovsky, Russian entrepreneur. He bought Yukos, Russia's second-largest oil company, and modernized it along Western lines. The Russian government arrested him on fraud charges in October 2003.

Kenneth Lay, CEO of Enron Corporation, 1985–2001. Lay, a close associate of President George W. Bush and major contributor to Bush's 2000 presidential campaign, was said to be one of the advisers to the task force that formulated the Bush administration's energy policy. Lay was indicted on fraud charges on July 7, 2004, but has pleaded innocent.

Amory and Hunter Lovins, an influential team who promote energy efficiency. Amory is a physicist and Hunter a lawyer and social scientist. In 1982, they cofounded the Rocky Mountain Institute, an independent nonprofit research center in Snowmass, Colorado. They rely on the private sector and market forces rather than government regulation to achieve their goals, and they encourage communication between businesses and the environmental movement. They were formerly married.

Nursultan Nazarbayev, president of Kazakhstan. He is accused of taking $78 million in bribes from New York merchant banker James Giffen on behalf of several oil companies.

Vladimir Putin, president of Russia, 1999–present. Some commentators have seen Putin's October 2003 arrest of oil company magnate Mikhael Khodorkovsky as a sign that Putin wants to take closer control of the Russian oil industry, possibly to the extent of renationalizing it.

Jeremy Rifkin, founder and president of the Foundation on Economic Trends. He claims that a combination of distributed electricity generation and widespread use of hydrogen fuel cells could produce a decentralized "hydrogen economy" that would free individuals and communities from a major source of control by governments and large corporations.

John D. Rockefeller, U.S. entrepreneur. He and his partners founded the Standard Oil Company of Ohio in 1870. He was famous for establishing a monopoly by controlling all aspects of the business of refining and distributing oil. The government broke up his Standard Oil Trust in 1892 and the Standard Oil Company of New Jersey, to which the trust had transferred its assets, in 1911, the latter following a Supreme Court decision that upheld the conviction of Rockefeller and the company for violation of the Sherman Antitrust Act. Rockefeller kept his fortune, however, and eventually donated much of it to philanthropic causes.

Franklin D. Roosevelt, president of the United States, 1932–45. Roosevelt supported legislation that broke up public utility holding companies in 1935. In 1945, he paid a personal visit to Abd al-Aziz ibn Saud, the founder of Saudi Arabia, in order to establish close ties with that oil-producing country.

Biographical Listing

Abd al-Aziz ibn Saud, founder of Saudi Arabia, the world's largest oil-producing and exporting country. He established close ties with the United States in the 1930s and 1940s, to their mutual benefit.

Lewis Strauss, Atomic Energy Commission chairman in the 1950s. In a 1954 speech, he predicted that, in a generation or so, electricity produced in nuclear plants would be "too cheap to meter."

Jules Verne, French science fiction writer. In an 1874 book called *The Mysterious Island,* he described an economy that used water as a fuel, an idea similar to the "hydrogen economy" predicted by Jeremy Rifkin and some other visionaries today, who see the hydrogen being produced from water by electricity generated from renewable sources.

James Watt, Scottish inventor. In 1765, he created an engine that used steam at high pressure to move a piston up and down in a cylinder. This engine proved capable of powering many kinds of machines and helped to make the Industrial Revolution possible.

George Westinghouse, U.S. inventor and manufacturer. In the 1890s, he promoted the use of alternating current, which allowed electricity to be transmitted over long distances.

Boris Yeltsin, president of Russia, 1991–99. He privatized the country's oil industry in the early 1990s, allowing entrepreneurs such as Mikhael Khodorkovsky, later known as the "oligarchs," to take control of it.

CHAPTER 5

GLOSSARY

acid rain Precipitation containing dilute sulfuric and nitric acids. It is produced when sulfur dioxide and nitrogen oxides, released into the atmosphere by burning fossil fuels, combine with water. Environmentalists have blamed it for damage to the forests and the deaths of fish in lakes and rivers.

Alien Tort Claims Act (ATCA) A law passed in 1789 to allow federal courts to have jurisdiction in a small number of cases involving violations of international law, such as piracy. Beginning in 1980, it was applied to accusations of torture and genocide carried out by foreign governments and, later, to suits involving human rights abuses allegedly abetted by U.S. oil companies during projects in developing countries. It is also called the Alien Tort Statute (ATS).

alternating current (AC) Electric current that flows back and forth in a wire, rather than always flowing in the same direction. Unlike direct current, alternating current can be transmitted over long distances with minimal energy loss. Compare with **direct current.**

alternative energy sources Sources of energy other than those most commonly used, such as wind, solar, and hydrogen power. Most, but not all, alternative energy sources are relatively harmless to the environment ("green"), renewable, or both. *See also* **conventional energy sources, "green" energy sources,** and **renewable energy sources.**

anthracite The highest grade of coal, shiny black and almost pure carbon. It burns more cleanly than bituminous or lignite coal.

avoided cost The cost that a utility would incur in generating a given amount of energy in the least expensive and most efficient manner. The Public Utilities Regulatory Policy Act (PURPA) required utilities to buy electricity from qualifying generators at avoided cost.

barrel A unit for measuring oil. One barrel of oil holds 42 gallons.

biomass Plant and animal wastes used as fuel.

bitumen A tarry form of petroleum found in tar sands or oil sands.

Glossary

bituminous coal The middle grade of coal, containing more impurities than anthracite but less than lignite.

blowout A sudden release of oil or gas from a well, also called a gusher. Early drillers found gushers thrilling, but today drillers realize that blowouts waste tremendous amounts of oil and can cause great environmental damage, so they prevent them whenever possible.

breeder reactor A type of nuclear power plant that fissions uranium to produce plutonium, then fissions the plutonium to produce still more energy.

British thermal unit (Btu) The amount of energy needed to raise the temperature of one pound of water by 1 degree Fahrenheit.

cap-and-trade program A program that sets a limit on the total amount of pollution of a certain type that an industry, country, or group of countries can produce and allows members that produce less than their allotted share of that pollution to sell "emission credits" to companies that produce more than their share.

carbon dioxide (CO_2) A colorless, odorless gas given off by (among other things) the burning of fossil fuels. An increase in the amount of this gas in the atmosphere is generally believed to be a major contributor to the greenhouse effect. *See also* **greenhouse effect.**

carbon sink A feature of Earth, such as a forest or an ocean, that absorbs carbon dioxide from the atmosphere.

Carter Doctrine The policy statement, made by President Jimmy Carter in 1980, that the United States would be willing to use military force to prevent the takeover of the oil-rich Persian Gulf region by an outside power.

Clean Air Act A law, passed in its present form by Congress in 1970 and revised several times, which limits the amounts of certain substances that electric power plants, automobiles, and other sources of pollution may release into the air.

Climate Change Levy A tax on domestic energy use that Britain began imposing in 2002. Money from the tax is used to help the country reduce its carbon dioxide emissions in an effort to prevent global warming.

cogeneration The process of producing heat and electricity from a fuel simultaneously.

combined cycle gas-fired power plants Energy generation plants in which the waste heat from gas-fired turbines is recaptured in boilers and used to heat steam that powers other turbines. Most analysts consider them the most efficient and least polluting of conventional power plants.

conventional energy sources The sources of energy most commonly used today. Fossil fuels (coal, oil, and natural gas), nuclear power, and (usually) hydroelectric power are classified as conventional energy sources.

converter A device that changes direct electric current into alternating current or vice versa.

Corporate Average Fuel Economy Standards (CAFE Standards) Separate standards for minimum fuel efficiency (in miles per gallon) for passenger cars and light trucks, to be averaged over a vehicle manufacturer's whole line of models. Congress established the program in 1975, and it went into effect with the 1985 model year.

cracking A process by which refineries break down large hydrocarbon molecules into smaller ones.

criteria air pollutants Pollutants for which the Clean Air Act sets national limits. As of 2004, they are ground-level ozone, sulfur dioxide, nitrogen oxides, carbon monoxide, particulates, and volatile organic compounds.

crude oil *See* **petroleum.**

deregulation The process of reducing government regulation of energy industries in order to encourage free market competition. It is also called restructuring.

direct current Electric current that flows in only one direction in a wire. Homes and businesses must have electricity in this form, but it cannot easily be transmitted over long distances. Compare with **alternating current.**

dispatchability The ability of an energy source to provide electricity that can be counted upon to be available for sending out to meet demand at any given time.

distributed generation Generation of electricity in many small facilities at or near the spots where the electricity will be used, rather than in large central power plants.

distribution The local parceling out of natural gas or low-voltage electricity to individual homes and businesses. Compare with **transmission.**

economy of scale A situation in which it costs less per item to produce large numbers of an identical product than to make smaller numbers of several slightly different products or, alternatively, to fulfill a single regulation than to fulfill many similar regulations.

efficiency The ratio of the energy given out by an energy-producing device or installation to the energy put into it.

electrolysis The process of using electricity to split water molecules into hydrogen and oxygen, a possible source of hydrogen to be used for energy.

energy The ability to do work—that is, to cause changes in something. Physical changes, such as melting of ice, and chemical reactions, as well as movement, require energy. Types of energy include electrical, chemical, nuclear, and thermal (heat).

energy merchant *See* **power marketer.**

Energy Policy Act (EPAct) A massive set of laws passed in 1992 that, among other things, broadened FERC's authority to order wheeling and exempted certain wholesale generators of electricity from regulation under PUHCA.

ethanol An alcohol that can be used as a fuel, either alone or combined with gasoline (gasohol), in standard internal-combustion automobile engines with minor modifications. It can be made from biomass and is less polluting than gasoline.

externalities Hard-to-define costs not included in figuring the price of a commodity, such as the social and environmental damage that fossil fuel extraction and use may cause. When externalities are not included, the price charged is lower than the commodity's true cost, which can lead to overconsumption.

Federal Energy Regulatory Commission (FERC) The independent federal government agency that regulates interstate shipping of oil, natural gas, and electricity and wholesale (but not retail) sales of gas and electricity; it replaced the Federal Power Commission in 1977.

Federal Power Act Part 1 of this act, passed by Congress in 1920, established the Federal Power Commission and gave it control of government hydroelectric projects. Congress passed part 2 of the act in 1935, extending the commission's authority to wholesale electric utility rates.

Federal Power Commission The ancestor of today's Federal Energy Regulatory Commission. Congress established it in 1920 with part 1 of the Federal Power Act and gave it control of government hydroelectric projects. Its regulatory power was extended to wholesale electric utility rates in 1935, rates for interstate pipeline transmission of natural gas in 1938, and wellhead prices for natural gas shipped between states in 1954.

Federal Surface Mining Control and Reclamation Act A law passed in 1977 that requires surface ("strip") mining companies to restore topsoil and vegetation to areas they have finished using.

fission The splitting apart of the nucleus of a heavy element, such as the radioactive isotope of uranium (U-235), by a free neutron; the process releases energy as well as more neutrons and can be used to produce heat that, in turn, generates electricity.

Foreign Corrupt Practices Act A law passed in 1977 that forbids U.S. citizens or corporations to pay bribes to foreign officials to obtain or retain business.

forum non conveniens A legal doctrine, meaning "inconvenient venue," stating that a trial must be held in the location most convenient to all parties.

fossil fuels Oil, coal, and natural gas, so called because most scientists believe they were made from the bodies of living things that died millions of years ago.

fuel cells Devices in which hydrogen combines with oxygen in a chemical reaction that produces water and energy (heat or electricity). Like batteries, they can be a way of storing electricity.

gasohol A combination of ethanol and gasoline that can be used as an automobile fuel. It produces less pollution than does gasoline.

geothermal energy Energy derived from heat inside the Earth.

global warming *See* **greenhouse effect.**

"green" energy sources Sources of energy whose extraction and use cause relatively little damage to the environment. Wind and solar power are examples.

greenhouse effect A rise in the temperature of the Earth's surface caused by increased levels of carbon dioxide and certain other gases in the atmosphere. These gases allow solar energy to reach Earth but limit the amount that reradiates into space, trapping heat energy near the planet's surface.

greenhouse gases Atmospheric gases believed to contribute to the greenhouse effect, including carbon dioxide and methane.

grid The network of transmission lines that carries high-voltage electricity across long distances from generators to distributors.

gusher *See* **blowout.**

head The height difference between the reservoir at the top of a dam and the basin below. Dams with high heads produce more energy from a given volume of water than those with low heads but also require larger reservoirs and cause greater environmental damage.

holding company A company formed to control other companies by owning large shares of their stock.

hydrates Compounds in which a substance is bound to water. Some scientists think that icelike methane hydrates in permafrost and the deep sea could become an inexhaustible source of natural gas.

hydrocarbon A compound of carbon and hydrogen. Fossil fuels consist almost entirely of hydrocarbon molecules, which can range from small and simple to large and extremely complex.

hydroelectric power Energy from moving water, usually water falling from a height through a dam to turn turbines and generate electricity. It is also called hydropower or hydro.

hydrogen The most abundant element in the universe. It is considered a promising energy source.

hydrogen economy A possible future economy described by Jeremy Rifkin and others, in which individuals and communities will use hydrogen fuel cells to generate most of their own electricity and power motor vehicles.

hydrothermal reservoir A place where superheated steam, hot water, or both, heated by geothermal energy, lie close to the Earth's surface.

Independent System Operator (ISO) A nonprofit organization, not affiliated with any utility or power company, that manages an electricity transmission network. It is one form of regional transmission organization (RTO). *See also* **regional transmission organization; transco.**

kerogen Organic matter in oil shale that potentially can be converted to oil.

kerosene A hydrocarbon that can be extracted from oil (petroleum) or coal by heating. It was a popular lamp fuel in the late 19th century.

Kyoto Protocol An agreement signed by representatives of about 180 countries in Kyoto, Japan, in December 1997. In order to prevent or limit global warming, it sets goals for reduction of carbon dioxide emissions by developed countries and establishes a cap-and-trade program for CO_2 emissions. It becomes international law in 2005, following Russia's ratification in late 2004.

lignite The lowest grade of coal, sometimes called "brown coal." It contains more impurities and therefore releases less energy and more pollutants than anthracite or bituminous coal.

market regulator *See* **swing producer.**

methane A simple hydrocarbon molecule made up of one carbon atom surrounded by four hydrogen atoms; it is the chief compound in natural gas.

methanol An alcohol that can be used as an automobile fuel. It produces less pollution than gasoline and can be made from biomass.

methyl tertiary butyl ether (MTBE) An oxygen-containing compound added to gasoline in some refineries to reduce carbon monoxide pollution as required by the Clean Air Act revision of 1990. Around the end of the 1990s, MTBE was found to be polluting groundwater and potentially harming human health. It is therefore being phased out.

Mineral Leasing Act A law passed in 1920 to establish rules for leasing of federal lands for oil and gas extraction.

monopoly An exclusive legal or de facto right to provide a good or service within a particular area, or a business that has such a right. *See also* **natural monopoly.**

National Energy Act A massive set of laws proposed by President Jimmy Carter to strengthen U.S. energy security by encouraging conservation and diversification of energy sources. Passed in 1978, it included both the Public Utility Regulatory Policies Act (PURPA) and the Natural Gas Policy Act.

natural gas A gas, consisting mostly of methane, that is used as a fuel. It is a fossil fuel, but it produces fewer pollutants (including carbon dioxide) than oil or coal.

Natural Gas Act Passed in 1938, this federal law gave the Federal Power Commission (later the Federal Energy Regulatory Commission) the right to regulate rates charged for interstate transmission of natural gas. Relatively unchanged, it still governs interstate natural gas transmission.

Natural Gas Policy Act This law, passed in 1978 as part of the National Energy Act, began the deregulation or restructuring of the natural gas industry by reducing government control of wellhead prices.

Natural Gas Wellhead Decontrol Act Passed in 1989, this law ended all federal caps on natural gas wellhead prices as of 1993.

natural monopoly An industry or part of an industry in which a single firm can provide a good or service more efficiently than two or more competing firms; for instance, building several competing networks of gas pipelines or electric transmission lines in the same area would be less efficient than building a single one. Rather than attempting to break up natural monopolies, governments regulate them to prevent them from charging unfairly high prices.

net metering A process in which consumers of electricity can generate part of their own electricity and sell any excess power back to their local utility, thus in effect making their electric meters "run backward."

new source review A process required by the Clean Air Act, whereby new or remodeled electric power plants, but not old ones, are required to meet certain limits on pollution.

NIMBY syndrome ("not in my back yard" syndrome) Behavior of people who theoretically approve of building facilities such as electric power plants but block such construction in their own neighborhoods.

nitrogen oxides Pollutants produced by burning fossil fuels; they contribute to smog and acid rain.

nuclear power Energy released by the splitting (fission) of the atomic nuclei of certain heavy elements, usually the radioactive form of uranium (U-235), which can be used to generate electricity.

Nuclear Regulatory Commission The government agency that replaced the Atomic Energy Commission as the chief regulator of nuclear power plants in 1974.

Nuclear Waste Policy Act A law passed in 1972 that authorizes creation of a national program to dispose of highly radioactive waste from nuclear power plants.

Oil Pollution Act A law passed in 1990, soon after the *Exxon Valdez* spill, that provides measures for prevention and cleaning up of oil spills.

oil sands *See* **tar sands.**

oil shale A type of rock from which oil may possibly be extracted with new techniques.

ozone A form of oxygen (O_3) that exists as a pale blue, strong-smelling gas. Ozone high in the atmosphere protects Earth against the Sun's ultraviolet radiation, but ground-level ozone is a major ingredient in smog and is one of the six "criteria pollutants" limited under the Clean Air Act.

Organization of Petroleum-Exporting Countries (OPEC) An organization formed by Saudi Arabia, Venezuela, Iran, Iraq, and Kuwait in the 1960s to coordinate policies to "secure fair and stable prices" for their oil. OPEC had 11 member countries in 2004.

oxygenating compounds Oxygen-containing compounds such as methyl tertiary butyl ether (MTBE) and ethanol, which the Clean Air Act requires to be added to gasoline in some places during the winter to reduce carbon monoxide pollution.

particulates Tiny particles of soot and ash that enter the atmosphere when certain fuels burn. They can help to cause respiratory disease.

peat Compacted masses of plant matter formed beneath marshes and lakes. Peat is a precursor of coal.

petroleum A blend of hydrocarbons in liquid form, commonly called crude oil. It is a fossil fuel.

photosynthesis The chemical process by which green plants transform carbon dioxide and water into compounds that store and use chemical energy within their bodies.

photovoltaic cell (PV cell) A small device, usually consisting of layers of silicon with certain added impurities, used to convert solar energy directly into electricity. It is also called a **solar cell.**

plasma A superhot, gaslike material consisting only of protons. Fusion reactions take place in a plasma.

plutonium A radioactive element that, like uranium, can be a fuel in nuclear power plants. So-called breeder reactors make plutonium from uranium and then fission the plutonium.

polymers Long, complex hydrocarbon molecules that can be refined from crude oil. They are used to make plastics, medicines, and many other substances.

positron A subatomic particle that is like an electron except that it has a positive instead of a negative electrical charge. Positrons are produced in fusion reactions.

potential energy Energy that is stored rather than currently doing work; an example is the energy in water behind a dam, which could do work when the dam is opened.

power broker A company that does not own, buy, or sell energy resources directly but acts as an agent to bring together buyers and sellers of such resources. Compare with **power marketer.**

power marketer A company that buys and sells energy on the open market but does not own or operate any energy facilities. Enron was an example. It is also referred to as an **energy merchant.** Compare with **power broker.**

primary energy source A source of energy that can be obtained from nature with little or no need for use of other forms of energy to transform it. Examples are wind, solar energy, coal, oil, and natural gas.

produced water Water extracted with oil and left as a waste product of the drilling process. It may contain toxic compounds.

proven reserves Quantities of unextracted oil or gas known to exist and believed to be removable with present technology, under current economic conditions.

public utility A private business that provides an essential service to the public and is given a monopoly on that service in a certain area. In return for this privilege, it is subject to government regulation to prevent price gouging and other abuses.

Public Utility Act A set of laws passed in 1935 that included the Federal Power Act and the Public Utility Holding Company Act.

public utility commission A body established by a state to regulate public utilities, such as gas and electricity companies, within the state. It usually assigns territories, sets service standards, regulates rates, and ensures that utilities serve all interested customers. In some states, this body is called a public service commission.

Public Utility Holding Company Act (PUHCA) A law passed by Congress in 1935 to regulate ownership and stock holdings of public utilities in order to combat perceived monopoly control and abuses by holding companies that owned many utilities at the time.

Public Utility Regulatory Policies Act (PURPA) A law passed as part of the National Energy Act in 1978, which required electric utilities to buy power offered by "qualifying facilities" (QFs), which were cogenerators and small nonutility generators that used renewable energy sources.

qualifying facilities (QFs) Electricity generating facilities from whom, according to the 1978 Public Utility Regulatory Policies Act (PURPA), utilities must buy power when offered. QFs were cogenerators or small nonutility generators that obtained their electricity from renewable sources.

refinery A plant in which physical and chemical processes are used to break crude oil (petroleum) into different kinds of hydrocarbon molecules.

reformer A device for extracting hydrogen from hydrocarbons for use in fuel cells.

regional transmission organization (RTO) An independent organization that manages the transmission of electricity within a multistate region. It may be either an Independent System Operator or a transco. *See also* **Independent System Operator** and **transco.**

renewable energy sources Energy sources that are constantly replenished, such as sunlight and wind. Most alternative energy sources, but not all, are renewable.

renewable portfolio standard A requirement by a public utility commission that generators or sellers of electricity obtain a specified percentage of their energy from renewable sources.

Renewables Obligation A combination of renewable portfolio standard and trading program that Great Britain launched in 2000 to promote use of renewable energy sources.

restructuring *See* **deregulation.**

secondary energy source A source of energy that must be transformed from primary sources, requiring input of additional energy, before it can be controlled and used. Electricity is the most common example.

sequestration The process of keeping carbon dioxide out of the atmosphere so it will not contribute to global warming. Possible sequestration methods include burying the gas and binding it in carbonate minerals.

seismic exploration A method of looking for oil and coal deposits that involves sending powerful sound waves into earth or water and analyzing the echoes generated when the waves bounce off the boundaries of layers of different kinds of rocks.

"Seven Sisters" A term used in the 1970s for the then-largest multinational oil companies: Royal Dutch/Shell, Exxon, Gulf, Texaco, BP, Mobil, and Standard Oil of California (Chevron).

Sherman Antitrust Act A law passed in 1890 to forbid monopolies, trusts, and other business arrangements or contracts that restrain trade.

solar cell *See* **photovoltaic cell.**

solar collector A device usually installed in groups on the roof of a building to collect energy from sunlight and use it to heat water or another fluid. It is basically a glass-covered box containing a black-painted metal plate connected to tubes filled with water. It also is called a solar panel.

spot market Market in which a commodity such as natural gas or electricity is traded at prices that change constantly as supply and demand vary.

Standard Market Design A set of rules that FERC recommended in 2002 that all state public utility commissions use in designing regulations.

stranded costs Costs for capital investments, such as building power plants, or long-term contracts with power generators that utilities were guaranteed to recover under traditional rate regulation but might not recover under competition.

Strategic Petroleum Reserve A stockpile of hundreds of millions of barrels of oil, owned by the U.S. government and kept in underground caves along the coast of Texas and Louisiana. It was established in 1975, after Arab countries cut off a substantial share of the oil imported by the United States. Its chief purpose is to guarantee a secure oil supply for the country in case of future cutoffs of imported oil.

strip mining Popular name for surface mining, a form of coal mining in which the land above shallow coal deposits is removed so the coal can be scooped out of the exposed surface.

sulfur dioxide (SO_2) A pollutant released by burning coal and other fossil fuels that contributes to acid rain.

sustainable development Development and use of natural resources in a way that preserves them for continued use by future generations.

swing producer A player in a market that can change overall prices by changing the amount it produces of the commodity being sold. Saudi Arabia is the world oil market's swing producer. Also called a market regulator.

system benefit charge A small charge levied on electricity consumers, proportional to the amount of energy they use; the money is used to support renewable energy or energy efficiency projects.

take-or-pay contract A contract that obliges a purchaser, such as a utility, pipeline company, or distributor, to pay a supplier for a certain minimum amount of product (usually natural gas or electricity) whether the purchaser actually takes that amount or not, in return for being guaranteed access to supply.

tar sands Also called oil sands, a geological formation in which sand grains are coated with envelopes of water and bitumen, a tarry form of petroleum. Extraction of oil from tar sands is beginning to become practical.

thermodynamics The science of energy. The transformation of energy is governed by two so-called laws of thermodynamics. The first law states that energy can be neither created nor destroyed. The second states that in a cyclic process, energy can never be changed completely into work; some energy is always converted into heat in the body that does the work.

tokamak A type of atomic fusion reactor that uses a magnetic field to contain the plasma in which the reaction takes place.

Torture Victim Protection Act A law passed in 1991, confirming that torture committed by foreign governments is a crime under U.S. law and allowing suits alleging torture to be brought under the Alien Tort Claims Act of 1789. *See also* **Alien Tort Claims Act.**

transco An independent, for-profit transmission company that manages the flow of electricity within a region; it is one possible form of regional transmission organization. *See also* **Independent System Operator** and **regional transmission organization.**

transformer A device that increases or decreases the voltage of an alternating electric current. Transformers are used to "step up" current from a generating station to a high voltage for efficient transmission and again to "step down" voltage for safe use once the electricity has reached its distribution area.

transmission Bulk movement of natural gas or high-voltage electricity over long distances. Compare with **distribution.**

turbine A device consisting of blades on the end of a shaft. When energy from pressurized steam or other sources pushes against and turns the blades, the shaft is made to rotate as well. The shaft can be attached to a loop of wire in a magnetic field and make the wire turn, which causes an electric current to flow back and forth in the wire (alternating current).

unbundling Requiring a vertically integrated utility to separate its producing, transmitting, and distributing functions.

vertical integration A business arrangement whereby a single company controls most or all aspects of industrial production; for example, an electricity utility might own generating plants, long-distance transmission lines, and facilities that distribute power to individual customers. Vertical integration improves coordination of activities and leads to economies of scale, but it can also result in undesirable monopoly control.

wheeling Transmission of electricity from a generator to a distributor over transmission lines owned by another utility.

PART II

GUIDE TO FURTHER RESEARCH

CHAPTER 6

HOW TO RESEARCH ENERGY ISSUES

Issues related to energy have generated a considerable amount of information in recent years. This chapter presents a selection of resources, techniques, and research suggestions for investigating energy supply, different types of energy sources, domestic and foreign policy issues, and other topics related to energy.

Although students, teachers, journalists, and other investigators may ultimately have different objectives, all are likely to begin with the same basic steps. The following general approach should be suitable for most purposes:

- Gain a general orientation by reading the first part of this book. Chapter 1 can be read as a narrative, while chapters 2–5 are best skimmed to get an idea of what is covered. They can then be used as a reference source for helping make sense of the events and issues encountered in subsequent reading.
- Skim some of the general books listed in the first section of the bibliography (chapter 7). Neutral overviews and books that provide pro and con essays on various issues in the field are particularly recommended.
- Browse the many web sites provided by organizations involved in energy issues (see chapter 8), including both neutral ones and advocacy groups that support or criticize particular energy sources or energy policies. Their pages are rich in news, articles, and links to other organizations, as well as describing particular cases and discussing the pros and cons of various aspects of energy generation and use.
- Use the relevant sections of chapter 7 to find more books, articles, and online publications on particular topics of interest.
- Find more (and more recent) materials by using the bibliographic tools such as the library catalogs and periodical indexes discussed later.

- To keep up with current events and breaking news, check back periodically with media and organization web sites and periodically search the catalogs and indexes for recent material.

The rest of this chapter is organized according to the various types of resources and tools. The three major categories are online resources, print resources, and the special area of law, legislation, and legal research.

ONLINE RESOURCES

With the increasing amount of information being made available online, turning to the World Wide Web is a logical way to begin any research project. It is easy to drown in the sea of information the Web reveals, however. Fortunately, starting with a few well-organized, resource-rich sites and then applying selective Web searching can provide a logical thread through the labyrinth.

GENERAL AND GOVERNMENT WEB SITES

As chapter 8 shows, dozens of groups present information or take stands on various energy-related issues. The following major sites are recommended as good starting places for research. They offer well-organized overviews of issues, provide numerous resources and links, and answer frequently asked questions. Most are government sites or sites of large intergovernmental organizations such as the International Energy Agency. Note that the U.S. Department of Energy has several useful sites.

- U.S. Department of Energy, URL: http://www.energy.gov/engine/content.do?BT_CODE=DOEHOME: DOE's main web site includes sections on national security, energy sources, energy efficiency, environment, energy prices and trends, science and technology, health, and safety and security, as well as news releases and information for teachers and young people.
- Energy Information Administration (U.S. Department of Energy), URL: http://www.eia.doe.gov/: One-stop shop for a tremendous quantity of statistics and reports on energy resources and use in the United States, other individual countries, and worldwide, accessible by country and by type of fuel. Includes historical data and projections to 2020. The EIA also has a "kids' page" at http://www.eia.doe.gov/kids/. The format of this page makes it seem aimed at elementary school, but it contains some information on technologies that is complex enough for high school.

- National Renewable Energy Laboratory (U.S. Department of Energy), URL: http://www.nrel.gov/: Site includes educational materials and research reports on all forms of renewable energy.
- Office of Energy Efficiency and Renewable Energy (U.S. Department of Energy), URL: http://www.eere.energy.gov/: This site advertises an "Energy Information Portal" that is "a gateway to hundreds of web sites and thousands of online documents on energy efficiency and renewable energy." It also describes a number of federal government programs related to these topics and provides numerous reports for downloading.
- Office of Fossil Energy (U.S. Department of Energy), URL:http://www.fe.doe.gov/: This site has material on coal and natural-gas power systems, carbon sequestration, hydrogen and other clean fuels, oil and gas supply and delivery, natural gas regulation, electricity regulation, and petroleum reserves. It includes many short fact sheets and descriptions of government programs in these areas.
- U.S. Environmental Protection Agency, URL: http://www.epa.gov/: Energy-related topics on this site include acid rain, the Clean Air Act, global warming, and oil spills.
- Energy Intelligence Group, URL: http://www.energyintel.com: Sells information to the global energy business, but the web site also provides free statistics and special reports on such subjects as global reliance on Middle Eastern oil.
- Energy Quest, URL: http://www.EnergyQuest.ca.gov: California Energy Commission's educational web site, written at about middle school/junior high school level. The site includes an introduction to energy, its sources and uses (The Energy Story), science projects, scientists associated with energy, games, links, library, and many types of resources for students and teachers.
- International Energy Agency, URL: http://www.iea.org: Includes free downloadable publications on numerous energy-related topics, from fusion power to sustainable development.
- World Energy Council, URL: http://www.worldenergy.org/wec-geis/: Site includes detailed statistical information on worldwide energy sources and use, market factors, policy, and technology, as well as news items about energy.

ADVOCACY WEB SITES

The following sites deal with and take a position on specific energy issues, especially the use of conventional as opposed to alternative/renewable fuels.

Some are sponsored by trade organizations or support groups for a particular type of fuel, such as petroleum or wind power, while others simply support a general view that favors, say, care for the environment and promotion of energy efficiency and renewable resources.

Conventional Fuels

- American Nuclear Society, URL: http://www.ans.org/pi: The society's Public Information page includes news items, position statements, resources, press releases, and material for teachers and students, as well as a link to another page that provides information about nuclear science and technology and ways it affects people's lives.
- American Petroleum Institute, Energy Consumers site, URL: http://apiec.api.org/frontpage.cfm: Petroleum and natural-gas industry trade organization site includes basic information about oil and natural gas, opinions on policy and environmental issues, descriptions of technical innovations in the field, news releases, educational materials, industry statistics, and more.
- Americans for Balanced Energy Choices, URL: http://www.balancedenergy.org/abec: Promotes the use of coal for generation of electricity. The site explains why coal power is "essential, affordable, and increasingly clean."
- Electric Power Supply Association, URL: http://www.epsa.org/forms/documents/DocumentFormPublic/: Trade association of competitive generators, power marketers, and other energy suppliers stresses the importance of competition in electricity generation. The site includes position papers and news items.
- Nuclear Energy Institute, URL: http://www.nei.org/: Explains the advantages of nuclear power as an energy source. Includes a page for students, news items, library, statistical data, and more.
- World Petroleum Congress, URL: http://www.world-petroleum.org/: Site includes educational material, news stories, and reports. It provides a useful global perspective.

Alternative/Renewable Fuels, Efficiency, and Environment

- American Wind Energy Association, URL: http://www.awea.org/: Site of trade association for wind power features a tutorial that provides "everything you wanted to know about wind energy" as well as news stories and publications.
- Consumer Energy Center, URL: http://www.ConsumerEnergyCenter.org: This California Energy Commission site provides consumer tips on

ways to increase energy efficiency in homes and businesses, information about alternative fuel vehicles and buying an energy efficient vehicle, and information about renewable energy.

- Edugreen, URL: http://www.edugreen.teri.res.in/: Site for young people (elementary/middle school) includes basic facts on energy, types of energy sources, and several related topics, including climate change and air pollution.
- HyWeb, URL: http://www.hydrogen.org/index-e.html: Provides information about hydrogen and hydrogen fuel cells, including news, companies and products, and political aspects.
- National Hydropower Association, URL: http://www.hydro.org/: Includes news stories, position papers, and other publications.
- Resources for the Future, URL: http://www.rff.org/: Scholarly site has general environmental emphasis, with discussions on climate change, renewable energy sources, and more.
- Rocky Mountain Institute, URL: http://www.rmi.org/: Founded by Amory and Hunter Lovins, this institute stresses energy efficiency, which it claims can simultaneously increase business profitability and improve the environment. Its web site has extensive resources for various audiences and discusses topics including buildings and land, businesses, climate, and transportation.
- Solar Energy Industries Association, URL: http://www.seia.org/: Trade organization web site includes basic information about solar energy, solar projects and equipment, and news stories.
- Worldwatch Institute, URL: http://www.worldwatch.org/topics/energy: Has an extensive research library on energy topics at this URL, including climate change, energy sources, materials, and transportation. The site has an overall environmental emphasis.

MEDIA SITES

News (wire) services, most newspapers, and many magazines have web sites that include breaking news stories and links to additional information. Many require fees to download articles. The following media sites have substantial listings for stories on energy topics.

- **Cable News Network (CNN):** http://www.cnn.com
- ***New York Times:*** http://www.nytimes.com (offers only abstracts unless you pay)
- **Reuters:** http://www.reuters.com

- ***Time* magazine:** http://www.time.com/time/ (offers only abstracts unless you pay, except for articles printed during the preceding week)
- ***Wall Street Journal:*** http://online.wsj.com/public/us (requires subscription to access articles)
- ***Washington Post:*** http://www.washingtonpost.com/ (requires free registration)

Yahoo! also maintains a large set of links to many newspapers that have web sites or online editions at http://dir.yahoo.com/News_and_Media/Newspapers/Web_Directories/.

FINDING MORE ON THE WEB

Although the resource sites mentioned earlier provide a convenient way to view a wide variety of information, the researcher will eventually want to seek additional data or views elsewhere. The two main approaches to Web research are the portal and the search engine.

WEB PORTALS

A Web guide or index is a site that offers what amounts to a structured, hierarchical outline of subject areas. This enables the researcher to zero in on a particular aspect of a subject and find links to web sites for further exploration. The links are constantly being compiled and updated by a staff of researchers.

The best known (and largest) Web index is Yahoo! (http://www.yahoo.com). Yahoo!'s home page gives a top-level list of topics, and the researcher simply clicks to follow them down to more specific areas. Alternatively, there is a search box into which the researcher can type one or more keywords and receive a list of matching categories and sites. (The box is rather confusingly labeled "Search the Web," but it also searches Yahoo!'s directories, and the results of this search appear at the top of the page.)

Web indexes such as Yahoo! have two major advantages over undirected "Web surfing." First, the structured hierarchy of topics makes it easy to find a particular topic or subtopic and then explore its links. Second, Yahoo! does not make an attempt to compile every link on the Internet (a task that is virtually impossible, given the size of the Web). Instead, Yahoo!'s indexers evaluate sites for usefulness and quality, which gives the researcher a better chance of finding more substantial and accurate information. The disadvantage of Web indexes is the flip side of their selectivity: The researcher is dependent on the indexer's judgment for determining what sites are worth exploring.

To research energy topics via Yahoo!, the researcher should browse to "Science—Energy" in the Yahoo! Web Directory. At the time of writing, the following topics appeared under this heading:

- Alternative Fuel Vehicles
- Biodiesel
- Biomass
- Companies
- Conservation and Efficiency
- Electrical Power Engineering
- Environmental Impact
- Events
- Flywheels
- Free Energy
- Fuel Cells
- Fusion
- Geothermal
- Government Agencies
- Hydropower
- Institutes
- Journals
- Methane Digester
- News and Media
- Nuclear
- Organizations
- Petroleum
- Policy
- Renewable
- Software
- Solar Power
- Statistics
- Wind Power

A variety of sites selected by the editors are also available for browsing. Some other topics, such as "Society—Environment and Nature", might lead to energy-related sites as well.

The Mining Company's About.com (http://www.about.com) is rather similar to Yahoo! but gives a greater emphasis to overviews or guides prepared by experts in various topics. The site does a good job of creating a guide page "on the fly" when a key word or phrase is entered in the search box. At present, "Power/Energy" is a subhead under "Industry & Business". In addition to specific news stories on energy topics, the "Power/Energy" page contains links to "Essentials," including Alaska oil, Iraq oil, and the Russian oil and gas sector, and to "Articles and Resources," including fuels, alternative energy, oil and gas production, grids/generation, nuclear power, coal energy, and hydropower. Note that About.com generates special URLs that keep pages "tied" to the About.com site, so for bookmarking purposes it is probably a good idea when visiting the linked site to reload it under its own URL. The site's own URL will be shorter than the About.com URL and is less likely to change.

New guide and index sites are constantly being developed, and capabilities are improving as the Web matures.

SEARCH ENGINES

Search engines take an approach to finding materials on the Web that is very different from that of web portals. Instead of organizing topically in a "top down" fashion, search engines work their way "from the bottom up," scanning through web documents and indexing them. There are hundreds of search engines, but some of the most widely used include:

- **AltaVista:** http://www.altavista.com
- **Excite:** http://www.excite.com
- **Google:** http://www.google.com
- **Hotbot:** http://www.hotbot.com
- **Lycos:** http://www.lycos.com
- **WebCrawler:** http://www.WebCrawler.com

To search with a search engine, one can employ the same sorts of keywords that work in library catalogs. There are a variety of Web-search tutorials available online (try "web search tutorial" in a search engine to find some). One good one is published by Bright Planet at http://www.brightplanet.com/deepcontent/tutorials/search/index.asp.

Here are a few basic rules for using search engines:

- When looking for something specific, use the most specific term or phrase. For example, when looking for information about *hydroelectric power*, use that specific term.
- Phrases should be put in quotes if you want them to be matched as phrases rather than as individual words. Examples might be *"Caspian Sea"*, *"electricity generation"*, and *"Alaska National Wildlife Refuge"*.
- When looking for a general topic that might be expressed using several different words or phrases, use several descriptive words (nouns are more reliable than verbs). For instance, *energy sources* is more likely to produce good results than *energy*, which often pulls up health-related topics as well as those related to power sources. Most engines will automatically put pages that match all terms first on the results list.
- Use "wildcards" when a desired word may have more than one ending. For example, *hydro** matches both *hydroelectric power* and *hydropower*. (Of course, it will also match *hydrogen* and numerous other words beginning with this prefix, not all of which may be energy-related.)
- Most search engines support Boolean *(and, or, not)* operators, which can be used to broaden or narrow a search:
- Use AND to narrow a search. For example, *electricity **and** generation* will match only pages that have both terms.
- Use OR to broaden a search: *alternative energy **or** renewable energy* will match any page that has either term, and since these terms are often used interchangeably, this type of search is necessary to retrieve the widest range of results.
- Use NOT to exclude unwanted results: *energy **not** health* should retrieve articles about energy but exclude those that use the term in a health sense.

Since each search engine indexes somewhat differently and offers somewhat different ways of searching, it is a good idea to use several search engines, especially for a general query. Some "metasearch" programs, such as Metacrawler (http://www.metacrawler.com) and SurfWax (http://www.surfwax.com), automate the process of submitting a query to multiple search engines. Metasearch engines may overwhelm you with results (and insufficiently prune duplicates), however, and they often don't use some of the more popular search engines, such as Google.

There are also search utilities that can be run from the researcher's own PC rather than through a web site. A good example is Copernic (http://www.copernic.com).

FINDING ORGANIZATIONS AND PEOPLE

Chapter 8 of this book provides a list of organizations involved with research or advocacy concerning energy issues. New organizations continue to emerge, however. The resource sites and web portals mentioned earlier are good places to look for information and links to organizations. If the name of an unfamiliar organization turns up during reading or browsing, the name can be given to a search engine. For best results, the complete name should be put in quotation marks (for instance, "*Alliance to Save Energy*"), although some search engines, such as Google, do not require this. If omitting the quotation marks, also omit common words such as *the* and *of*; for instance, type *alliance save energy* rather than the organization's complete name. Including these common words will confuse the search engine.

Another approach is to take a guess at the organization's likely web address. For example, the Organization of Petroleum Exporting Countries is best known by the acronym OPEC, so it is not a surprise that the organization's web site is at http://www.opec.org. (Note that noncommercial organization sites normally use the *.org* suffix, government agencies use *.gov*, educational institutions have *.edu*, and businesses use *.com*.) This technique can save time, but it doesn't always work. In particular, watch out for "spoof" sites that mimic or parody organizational sites. Of course, parody sites may be of interest in themselves as forms of criticism or dissent.

When reading materials from an unfamiliar author, it is often useful to learn about that person's affiliation, credentials, and other achievements. There are several ways to find a person on the Internet:

- Put the person's name (in quotes) in a search engine, which may lead you to that person's home page or a biographical sketch put out by the institution for whom the person works.
- Contact the person's employer (such as a university for an academic, or a corporation for a technical professional). Most such organizations have web pages that include a searchable faculty or employee directory.
- Try a people-finder service, such as Yahoo! People Search (http://people.yahoo.com) or BigFoot (www.bigfoot.com). This may yield contact information, including an e-mail address, regular address, and/or phone number.

PRINT SOURCES

As useful as the Web is for quickly finding information and the latest news, in-depth research can still require trips to the library or bookstore. Getting the most out of the library, in turn, requires the use of bibliographic tools and resources. *Bibliographic resources* is a general term for catalogs, indexes, bibliographies, and other guides that identify the books, periodical articles, and other printed materials that deal with a particular subject. They are essential tools for the researcher.

LIBRARY CATALOGS

Most readers are probably familiar with the basics of using a library catalog, but they may not know that many catalogs besides that of their local library can be searched online. The largest library catalog, that of the Library of Congress, can be accessed at http://catalog.loc.gov, a page that includes a guide to using the catalog, as well as both basic and advanced catalog searches. Yahoo! offers a categorized listing of libraries at http://dir.yahoo.com/Reference/Libraries/.

Most catalogs can be searched in at least the following ways:

- An author search is most useful if you know or suspect that a person has written a number of works of interest. However, it may fail if you don't know the person's exact name. (Cross references are intended to deal with this problem, but they can't cover all possible variations.)
- A title search is best if you know the exact title of the book and just want to know if the library has it. Generally you need only use the first few words of the title, excluding initial articles *(a, an, the)*. This search will fail if you don't have the exact title.
- A keyword search will match words found anywhere in the title. It is thus broader and more flexible than a title search, although it may still fail if all keywords aren't present.
- A subject search will find all works to which the library has assigned that subject heading. The advantage of a subject search is that it doesn't depend on certain words being in a book's title. However, using this kind of search can require knowing the appropriate Library of Congress subject headings for a topic. You can obtain these from the Library of Congress catalog site (http://catalog.loc.gov) by clicking on "Basic Search", then selecting "Subject Browse" and typing in a term such as *oil industry*. On the next list that comes up, click subject headings of interest.

Once the record for a book or other item is found, it is a good idea to see what additional subject headings and name headings have been assigned to that item. These, in turn, can be used for further searching. For instance, in addition to *solar energy*, you will probably also want to check out *solar heating* and *solar air conditioning*.

BOOKSTORE CATALOGS

Many people have discovered that online bookstores such as Amazon.com (www.amazon.com) and Barnes & Noble (www.barnesandnoble.com) provide convenient ways to shop for books. A less-known benefit of online bookstore catalogs is that they often include publisher's information, book reviews, and readers' comments about a given title. They can thus serve as a form of annotated bibliography. Out-of-print or highly specialized materials may not appear in such catalogs, however.

BIBLIOGRAPHIES, INDEXES, AND DATABASES

Printed or online bibliographies provide a convenient way to find books, periodical articles, and other materials. Some bibliographies include abstracts (brief summaries of content), while others provide only citations. Some bibliographies and indexes are available online (at least for recent years), but you may be able to access them only through a library where you hold a card. (If you are on a college campus, ask a university reference librarian for help.) There are two good indexes with unrestricted search access, however. IngentaConnect (http://www.ingentaconnect.com) contains brief descriptions of about 13 million documents from about 27,000 journals in just about every subject area. Copies of complete documents can be ordered with a credit card, or they may be obtainable for free at a local library.

PERIODICAL INDEXES

Most public libraries subscribe to database services such as InfoTrac or EBSCO host, which index articles from hundreds, perhaps thousands, of general-interest and specialized periodicals. This kind of database can be searched by author or by words in the title, subject headings, and sometimes words found anywhere in the article text. Depending on the database used, "hits" can produce just a bibliographical citation (author, title, pages, periodical name, issue date, etc.), a citation and abstract, or the full text of the article. Before using such an index, it is a good idea to view the list of newspapers and magazines covered and determine the years of coverage.

Many libraries provide dial-in, Internet, or telnet access to their periodical databases as an option in their catalog menu. However, licensing restrictions usually mean that only researchers who have a library card for that particular library can access the database (by typing in their name and card number). Check with local public or school libraries to see what databases are available.

For periodicals not indexed by Infotrac or another index (or for which only abstracts rather than complete text is available), check to see whether the publication has its own web site (most now do). Some scholarly publications are putting most or all of their articles online. Popular publications tend to offer only a limited selection. Some publications of both types offer archives of several years' back issues that can be searched by author or keyword.

Nearly all newspapers now have web sites with current news and features. Generally, a newspaper offers recent articles (perhaps from the last 30 days) for free online access. Earlier material can often be found in an "Archive" section. A citation and perhaps an abstract is frequently available for free, but a fee of a few dollars may be charged for the complete article. One can sometimes buy a "pack" of articles at a discount as long as the articles are retrieved within a specified time. Of course, back issues of newspapers and magazines may also be available in hard copy, bound, or microfilm form at local libraries.

LEGAL RESEARCH

As with all complex and controversial topics, energy and related issues have been the subject of intense litigation in the courts. Human rights and environmental groups have used lawsuits to pressure multinational oil companies to take responsibility for alleged social and environmental damage resulting from their projects in developing countries, for example. Although one can find news coverage of some important cases in the general media, many researchers will need to find specific court opinions or the text of existing or pending legislation.

Because of the specialized terminology of the law, legal research can be more difficult to master than bibliographical or general research tools. Fortunately, the Internet has also come to the rescue in this area, offering a variety of ways to look up laws and court cases without having to pore through huge bound volumes in law libraries (which may not be easily accessible to the general public, anyway). To begin with, simply entering the name of a law, bill, or court case into a search engine will often lead the researcher directly to both text and commentary.

FINDING LAWS

Federal legislation is compiled into the massive U.S. Code. The U.S. Code can be searched online in several locations, but the easiest site to use is probably that of the Cornell Law School at http://assembler.law.cornell.edu/uscode/. The fastest way to retrieve a law is by its title and section citation (listed for all laws discussed in chapter 2), but popular names (Clean Air Act, Energy Policy Act, and so on) and keywords can also be used.

Many state agencies have home pages that can be accessed through the Findlaw state resources web site (http://findlaw.com/11stategov/). This site also has links to state law codes. These links may not provide access to the text of specific regulations, however.

KEEPING UP WITH LEGISLATIVE DEVELOPMENTS

Pending legislation is often tracked by advocacy groups, both national and those based in particular states. See Chapter 8, "Organizations and Agencies," for contact information.

The Library of Congress Thomas web site (http://thomas.loc.gov/) includes files summarizing legislation by the number of the Congress (each two-year session of Congress has a consecutive number: For example, the 109th Congress is in session in 2005 and 2006). Legislation can be searched for by the name of its sponsor(s), the bill number, or by topical keywords. (Laws that have been passed can be looked up under their Public Law number.) Only the current session can be searched from the main page, but earlier sessions can be searched by clicking on Search Bills and Resolutions under Legislation. Sessions back to the 101st (1989–90) can be accessed with Simple Search; summary and status information about bills introduced in previous sessions must be reached through Advanced Search. Further details retrievable by clicking on the bill number and then the link to the bill summary and status file include sponsors, committee action, and amendments.

A second extremely useful site is maintained by the Government Printing Office (http://www.gpoaccess.gov/index.html/). This site has links to the Code of Federal Regulations (which contains federal regulations that have been finalized), the Federal Register (which contains announcements of new federal agency regulations), the Congressional Record, the U.S. Code, Congressional bills, a catalog of U.S. government publications, and other databases. It also provides links to individual agencies, grouped under government branch (legislative, executive, judicial), and to regulatory agencies, administrative decisions, core documents of U.S. democracy such as the Constitution, and hosted federal web sites.

FINDING COURT DECISIONS

Legislation is only part of the story, of course. The Supreme Court and state courts make important decisions every year that determine how the laws are interpreted. Like laws, legal decisions are organized using a system of citations. The general form is: *Party1* v. *Party2 volume reporter* [optional start page] *(court, year)*.

Here is one example: *Standard Oil Company of New Jersey et al. v. United States*, 221 U.S. 1 (1911). Here the parties are Standard Oil Company of New Jersey (and others) and the United States (the first listed is the plaintiff or appellant, the second the defendant). The case is in volume 221 of the *U.S. Supreme Court Reports*, beginning on page 1, and the case was decided in 1911. (For the U.S. Supreme Court, the name of the court is omitted.)

A state court decision can generally be identified because it includes the state's name. For example, in *Aguinda et al. v. Texaco*, 142 F. Supp. 2d 534 (SDNY 2001), the 142 F. Supp. 2nd refers to the federal district court to which the case was transferred, but SDNY refers to the New York state court where it was first heard.

Once the jurisdiction for a case has been determined, the researcher can then go to a number of places on the Internet to find cases by citation and sometimes by the names of the parties or by subject keywords. Some of the most useful sites are:

- The Legal Information Institute (http://supct.law.cornell.edu/supct/) has all Supreme Court decisions since 1990, plus 610 of the most important historic decisions.
- Washlaw Web (http://www.washlaw.edu/) has a variety of courts (including states) and legal topics listed, making it a good jumping-off place for many sorts of legal research. However, the actual accessibility of state court opinions (and the formats they are provided in) varies widely.

LEXIS AND WESTLAW

Lexis and Westlaw are commercial legal databases that have extensive information, including an elaborate system of notes, legal subject headings, and ways to show relationships between cases. Unfortunately, these services are too expensive for most individual researchers to use unless they can access the services through a university or corporate library.

MORE HELP ON LEGAL RESEARCH

For more information on conducting legal research, see the "Legal Research FAQ" at http://www.faqs.org/faqs/law/research/. After a certain point, however, the researcher who lacks formal legal training may need to consult with or rely on the efforts of professional researchers or academics in the field.

A WORD OF CAUTION

Thanks to the Web, there is more information from more sources available than ever before. There is also a greater diversity of voices, since any person or group with a personal computer and Internet service can put up a web site—in some cases a site that looks as polished and professional as that of an established group. One benefit of this situation is that dissenting views can be found in abundance, including even sites maintained by extremist groups or their supporters. However, the other side of the coin is that the researcher—whether journalist, analyst, teacher, or student—must take extra care to try to verify facts and to understand the possible biases of each source. Some good questions to ask include:

- Who is responsible for this web site?
- What is the background or reputation of the person or group?
- Does the person or group have a stated objective or agenda?
- What biases might this person or group have?
- Do a number of high-quality sites link to this one?
- What is the source given for a particular fact? Does that source actually say what is quoted? Where did *they* get that information?

In a sense, in the age of the Internet each person must be his or her own journalist, verifying sources and evaluating the extent to which they can be relied upon.

CHAPTER 7

ANNOTATED BIBLIOGRAPHY

Thousands of books, articles, and Internet documents on energy and related issues have appeared in recent years. This bibliography lists a representative sample of material dealing with this subject. Sources have been selected for clarity and usefulness to the general reader, recent publication (most from 2000 or later), and a variety of points of view.

Listings are grouped in the following subject categories:

- general works about energy and energy policy
- conventional energy sources (oil, coal, natural gas, nuclear)
- alternative/renewable energy sources (hydropower, wind, solar, etc.) and energy efficiency and conservation
- utilities (electricity and natural gas) and their regulation
- global energy issues (oil-producing countries, international organizations, conflicts and wars, political and social effects)
- environmental issues (chiefly pollution and global warming)

Items are listed only once, under what appears to be the most important category, even though they might also fit under other categories.

Within each category, items are listed by type (books, articles, and web documents). Newspaper articles have not been included because magazines usually cover the same material, and back issues of magazines are easier to obtain than those of most newspapers. Magazine articles available on the Internet are listed as articles, not as web documents. Conversely, reports and similar material available for free downloading on the Internet are listed as web documents, even if they have also been printed at some time, because it is easier and less expensive to obtain them on the Internet than to send (and possibly pay) for paper copies.

Energy Supply

GENERAL WORKS ABOUT ENERGY AND ENERGY POLICY

BOOKS

Beaudreau, Bernard C. *Energy and the Rise and Fall of Political Economy.* Westport, Conn.: Greenwood Publishing Group, 1999. Studies the relationship between changes in energy use and changes in political economy during the 19th and 20th centuries.

Bent, Robert, Randall Baker, and Lloyd Orr, eds. *Energy: Science, Policy, and the Pursuit of Sustainability.* Washington, D.C.: Island Press, 2002. This introduction to interactions among energy, society, and the environment describes their physical, human, and political-economic dimensions.

Blackwood, John R., ed. *Energy Research at the Cutting Edge.* Hauppauge, N.Y.: Nova Science, 2002. Studies describe the current and future state of the energy industry, including consideration of both conventional and alternative fuels. Authors conclude that the conventional sources are more practical.

Borowitz, Sidney. *Farewell Fossil Fuels: Renewing America's Energy Policy.* Cambridge, Mass: Perseus Book Group, 1999. Describes major conventional and alternative sources of energy, considering the advantages and disadvantages of each.

Bosselman, Fred, Jim Rossi, and Jacqueline Lang Weaver. *Energy, Economics and the Environment: Cases and Materials.* New York: Foundation Press, 2000. Surveys U.S. laws and cases related to all types of energy sources and their use, as well as utility regulation and other energy issues such as global warming.

Boyle, Godfrey, Bob Everett, and Janet Ramage, eds. *Energy Systems and Sustainability: Power for a Sustainable Future.* New York: Oxford University Press, 2003. Introduces the policy, economic, social, and environmental issues raised by present energy use. The authors describe how the current energy picture evolved and consider how technological developments could ameliorate sustainability problems in the future.

Capello, Roberta, Peter Nijkamp, and Gerard Pepping. *Sustainable Cities and Energy Policies.* New York: Springer Verlag, 1999. Compares and evaluates energy policies of different European cities. The authors stress the usefulness of decentralized energy policies and renewable energy sources in achieving sustainable, environmentally benign cities.

Cleveland, Cutler J., ed. *Encyclopedia of Energy.* St. Louis, Mo.: Elsevier Science, 2004. Reference work written by leading authorities covers all fields of energy, including scientific, social, and environmental aspects.

Deutch, John, and Richard K. Lester, eds. *Making Technology Work: Applications in Energy and the Environment.* New York: Cambridge University

Press, 2003. Points out that successfully applying new technologies requires consideration of economic, political, social, and environmental factors as well as scientific and engineering ones.

Dupler, Douglas. *Energy: Is There Enough?* Detroit.: Information Plus, 2001. Reference book examines range of energy sources and their contribution to the world's energy supply.

Fitzgerald, Edward A. *The Seaweed Rebellion: Federal-State Conflicts over Offshore Energy Development.* Lanham, Md.: Rowman & Littlefield, 2000. Uses federal-state conflicts over offshore energy development from the 1930s through the 1990s as a way to examine the role of the courts in the public policy process. Fitzgerald claims that judicial favoring of federal over state interests harmed both the environment and coastal energy development.

Goldemberg, Jose, ed. *World Energy Assessment: Energy and the Challenge of Sustainability.* New York: United Nations Press, 2001. Analyzes economic, social, security, and environmental issues linked to energy supply and use. The book evaluates policy options in terms of sustainability and concludes that a sustainable future is possible if the right choices are made.

Helm, Dieter. *Energy, the State, and the Market.* New York: Oxford University Press, 2003. Describes British energy policy since 1979.

Hunt, Lester C., ed. *Energy in a Competitive Market: Essays in Honour of Colin Robinson.* Northampton, Mass.: Edward Elgar, 2003. Essays discuss the whole range of energy economics, with a focus on the question of whether markets are the key to effective allocation of energy resources.

International Energy Agency. *Energy: The Next Fifty Years.* Paris: Organisation for Economic Cooperation and Development, 2000. Papers from an OECD Forum for the Future conference predict what will happen in the energy sector in the next five decades, focusing on the technological, economic, environmental, and geopolitical challenges likely to be presented.

Mansfield, Marla E. *Energy Policy: The Reel World: Cases and Materials on Resources, Energy, and Environmental Law.* Durham, N.C.: Carolina Academic Press, 2001. Casebook covers all fields of U.S. energy law and regulation and their interaction with environmental law.

McVeigh, J. C., and J. G. Mordue, eds. *Energy Demand and Planning.* New York: Brunner-Routledge 1998. Essays discuss issues that will affect energy policy planning of the world energy industry during the next 50 years, including global warming, technological change, sustainable development, and population growth.

Plunkett, Jack W. *Plunkett's Energy Industry Almanac 2002–2003: The Only Complete Guide to the American Energy and Utilities Industry.* Houston, Tex.: Plunkett Research Ltd., 2001. Provides a complete overview of the energy industry, including profiles of the top petroleum-exporting countries and major energy companies.

Rifkin, Jeremy. *The Hydrogen Economy: The Creation of the Worldwide Energy Web and the Redistribution of Power on Earth.* New York: Jeremy P. Tarcher, 2003. Describes a utopian society in which individuals and communities gain political and economic independence through distributed (local) generation of electricity, powered by hydrogen obtained from water with energy from renewable sources.

Schobert, Harold. *Energy and Society: An Introduction.* Washington, D.C.: Taylor & Francis, 2001. Introduction to work, energy, and efficiency for the general reader describes history of energy use, different kinds of energy sources, energy-related technology, and environmental problems caused by energy use.

Shepherd, William, and David Shepherd. *Energy Studies*, 2nd ed. London: Imperial College Press, 2003. Reviews both conventional and alternative energy sources.

Simon, Andrew L. *Energy Resources.* Rochester, N.Y.: Simon Publications, 2001 (reprint). Introduces the various sources and carriers of energy, including their availability and the technologies used to produce them.

Smil, Vaclav. *Energy at the Crossroads: Global Perspectives and Uncertainties.* Cambridge: MIT Press, 2003. General survey of energy issues in the past, present, and future. Smil emphasizes economic, security, and environmental reasons for reducing dependency on fossil fuels.

Stevens, Paul, ed. *The Economics of Energy.* Northampton, Mass.: Edward Elgar, 2000. Two-volume set of reprints of the most significant research in the field during the past 70 years. Volume I focuses on demand-side issues, and Volume II deals with supply-side issues.

Vaitheeswaran, Vijay V. *Power to the People: How the Coming Energy Revolution Will Transform an Industry, Change Our Lives, and Maybe Even Save the Planet.* New York: Farrar, Straus & Giroux, 2003. Short essays cover energy problems from pollution to electricity deregulation. Author suggests that distributed generation and hydrogen fuel cell technology will solve many of them.

Wright, Russell O. *Chronology of Energy in the United States.* Jefferson, N.C.: McFarland & Co., 2003. Lists events in the history of energy use in the United States, stressing technological innovation.

ARTICLES

Baker, Chris, and Gary Anderson. "Power Hungry People." *Insight on the News*, vol. 17, June 25, 2001, p. 32. Energy use in the United States has been rising rapidly—and most of the energy comes from fossil fuels, leading to increased risk of wars fought over ownership of these scarce resources.

Brown, Matthew H., Christie Rewey, and Troy Gagliano. "Nine Hot Energy Issues." *State Legislatures*, vol. 30, April 2004, pp. 12–13. Describes nine energy issues important to state governments.

Carey, John. "Taming the Oil Beast." *Business Week*, February 24, 2003, pp. 96 ff. Recommends six steps to reduce dependence on oil, including diversifying sources of oil, using strategic reserves, increasing energy efficiency in industry and transportation, encouraging use of renewable sources, and phasing in fuel taxes.

Crook, Clive. "Read My Lips: There's No Energy Crisis." *National Journal*, vol. 33, May 12, 2001, p. 1383. U.S. energy policy needs to stop shielding consumers from price fluctuations.

Easterbrook, Gregg. "The Producers—How the Oilmen in the White House See the World." *New Republic*, June 4, 2001, pp. 27 ff. Both the George W. Bush administration and its opponents need to face some hard truths if a sound energy policy is to be crafted.

Ellwood, Wayne. "Mired in Crude." *New Internationalist*, June 2001, pp. 9–10. For economic, political, social, and environmental reasons, the world needs to begin the difficult process of weaning itself away from oil consumption as soon as possible.

Helm, Dieter. "It All Boils Down to Money." *New Statesman*, vol. 132, February 2, 2003, pp. xiv–xv. Britain faces serious challenges in meeting future energy goals.

Huber, Peter. "Gasoline and the Grid." *Forbes*, vol. 170, November 25, 2002, p. 154. Rather than trying to reduce demand for energy, federal policymakers should encourage substitution of electricity for oil.

Kriz, Margaret. "Energy and Politics." *Nieman Reports*, vol. 58, Summer 2004, pp. 10–11. A journalist describes the George W. Bush energy policy and energy stories that are likely to remain important in the near future.

Marshall, Carol. "Energy." *School Library Journal*, vol. 47, July 2001, p. S23. Lists of energy-related web sites for students.

Palmer, Mark J. "Oil and the Bush Administration." *Earth Island Journal*, vol. 17, Autumn 2002, pp. 20–22. Description and critique of the George W. Bush energy policy plan, which is alleged to strongly favor oil interests.

Rifkin, Jeremy. "Hydrogen: Empowering the People." *The Nation*, vol. 275, December 23, 2002, p. 20. By providing a decentralized but interconnected energy "web," hydrogen fuel cell technology has the potential to free society from centralized control.

Samuelson, Robert J. "Power Policy." *Newsweek*, May 22, 2001, p. 28. Conflicting public demands make deciding on an energy policy difficult.

"The Slumbering Giants Awake." *The Economist*, February 10, 2001, p. 2. The energy industry, including both utilities and oil, is changing rapidly. Shareholder value, convergence of companies, and risk are the factors that will shape the industry's future.

Smith, John F., Jr. "Technology for the New Millennium." *USA Today*, vol. 129, January 2001, pp. 26–27. Encouraging technological innovation is the only way to meet the conflicting demands of the market and the environment regarding energy.

Tanenbaum, B. Samuel. "Our Energy Appetite." *World and I*, vol. 17, August 2002, pp. 148–149. In the final quarter of the 20th century, the meteoric rise in energy use in the United States slowed down—but the slowdown may not last.

Warren, Andrew. "The Future Could Be Dark." *New Statesman*, vol. 130, June 18, 2001, p. 30. The author argues that Britain must make radical changes in its domestic energy policy to deal with climate change and other challenges.

Warshall, Peter. "The Unholy Triumvirate." *Whole Earth*, Winter 2001, pp. 32–36. Explains how energy, water, and money are inextricably intertwined.

Weidenbaum, Murray L. "Meeting the Global Energy Challenge." *USA Today*, vol. 130, January 2002, pp. 21–23. Developing a sound, consistent domestic energy policy with a strong market emphasis is necessary to protect the United States from global supply uncertainties.

Worthington, Barry. "We Still Have an Energy Crisis." *World and I*, vol. 17, August 2002, pp. 30–31. Supportive summary of the George W. Bush administration's energy plan.

Zuckerman, Mortimer B. "Speaking Truth About Energy." *U.S. News & World Report*, February 18, 2002, p. 68. Both conservation and new domestic drilling for oil will be necessary to reduce dependence on oil imports.

WEB DOCUMENTS

"The Bush-Cheney Energy Plan: How It Fares in the 21st Century." Energy Foundation. Available online. URL: http://www.ef.org/national/NationalAnalysis.pdf. Downloaded on July 22, 2004. Claims that the plan looks backward to conventional energy technologies and will hamper the U.S. economy.

"Challenge and Opportunity: Charting a New Energy Future." Energy Future Coalition. Available online. URL: http://www.energyfuturecoalition.org/files/pdf/report.pdf. Downloaded on July 22, 2004. Stresses reducing transportation's dependence on oil, reducing greenhouse gas emissions, and helping developing countries grow by extending modern energy services to them.

Annotated Bibliography

"The Energy Story." California Energy Commission, EnergyQuest. Available online. URL: http://www.energyquest.ca.gov/story/index.html. Downloaded on August 2, 2004. Guide for young people discusses the basic nature of energy, different sources of energy, and uses of energy such as electricity and transportation.

"International Energy Outlook 2004." Energy Information Administration. Available online. URL: http://www.eia.doe.gov/oaifieo. Posted on May 17, 2004. This government agency's predictions for the world energy picture in 2025.

"National Energy Policy Initiative Expert Group Report." National Energy Policy Initiative. Available online. URL: http://nepinitiative.org/pdfs/NEPInit_Report.pdf. Posted in 2002. Discusses transportation and mobility, electricity services, energy security, and climate change caused by energy use. The report calls for stronger government leadership and policy formation, as well as more support for research and development.

"National Energy Security Post-9/11." U.S. Energy Association. Available online. URL: http://www.usea.org/USEAReport.pdf. Posted in June 2002. Considers supply security, distribution/storage, and emergency preparedness/public safety issues for different energy sources; also discusses security issues concerning conservation, energy efficiency, and advanced energy technologies. The site predicts future scenarios and makes policy recommendations.

"Reliable, Affordable, and Environmentally Sound Energy for America's Future." National Energy Policy Development Group (White House web site). Available online. URL: http://www.whitehouse.gov/energy/. Posted in May 2001. Report of a task force headed by Vice President Dick Cheney proposes a national energy policy that stresses the need for increased domestic energy production.

"Toward a National Energy Strategy." U.S. Energy Association. Available online. URL: http://www.usea.org/NationalEnergyStrategy.pdf. Posted in February 2001. Makes policy recommendations for enhancing energy supplies, encouraging energy efficiency and affordable prices, stimulating global trade and international development, promoting energy technology development, balancing energy use and environmental concerns, and unifying the policy and regulatory process.

"United States of America." Energy Information Administration. Available online. URL: http://www.eia.doe.gov/emeu/cabs/usa.html. Posted in January 2005. Summary of U.S. energy production and consumption, including contributions of different energy sources, energy imports and exports, electricity, and environmental issues.

"Why Energy Matters." Energy Future Coalition. Available online. URL: http://www.whyenergymatters.org/get_guide/index.html. Downloaded on July 22, 2004. Discusses oil dependence, fuel economy, electric power, global warming, new technologies, and what consumers can do to reduce dependence on oil and prevent global warming.

"World Energy, Technology and Climate Policy Outlook 2030." European Commission. Available online. URL: http://europa.eu.int/comm/research/energy/pdf/weto_final_report.pdf. Posted in May 2003. Describes global energy challenges likely to become severe during the next 30 years, including the growing energy needs of developing countries and the need to cut greenhouse gases and shift to renewable energy sources.

CONVENTIONAL ENERGY SOURCES

BOOKS

Afgan, Naim Hamdia, and Maria da Gracia Carvalho, eds. *Natural Gas in Asia: The Challenges of Growth in India, China, Japan, and Korea.* New York: Oxford University Press, 2002. The next two decades will determine whether natural gas becomes an important fuel in these countries or remains peripheral.

Black, Brian. *Petrolia: The Landscape of America's First Oil Boom.* Baltimore, Md.: Johns Hopkins University Press, 2000. Describes the environmental and economic ruin of the area in northwestern Pennsylvania that was the site of the first oil boom in the United States in the 1860s.

Bryce, Robert. *Cronies: Oil, the Bushes, and the Rise of Texas, America's Superstate.* New York: Public Affairs, 2004. Stresses influence of Texas-based oil corporations and related firms such as Halliburton on U.S. and world politics, especially during both of the Bush administrations.

Campbell, C. J. *The Coming Oil Crisis.* Brentwood, Essex, U.K.: Multi-Science Publishing Co., Ltd., 2004. Eminent geologist claims that world crude oil production will begin to decline sharply during the first decade of the 21st century.

Clark, James A., and Michel T. Halbouty. *Spindletop: The True Story of the Oil Discovery that Changed the World.* Houston, Tex.: Gulf Publishing Co., 1999 (reprint). Describes the discovery of the first big oil field in Texas, which began producing with a giant gusher on January 10, 1901. The book was first published in 1952.

Cohen, Bernard Leonard. *The Nuclear Energy Option: An Alternative for the '90s.* Revised edition. Cambridge, Mass.: Perseus Press, 2000. Claims that nuclear power is the best choice for electricity generation because oil and coal produce byproducts that threaten human health as well as the environment, and alternative sources such as solar energy are impractical.

Deffeyes, Kenneth S. *Hubbert's Peak: The Impending World Oil Shortage.* Princeton, N.J.: Princeton University Press, 2003. Explains why the author believes that world oil production will peak in the next few years. The book also provides information on oil geology, exploration, and extraction.

Doern, G. Bruce, Aslan Dorman, and Robert W. Morrison, eds. *Canadian Nuclear Energy Policy: Changing Ideas, Institutions, and Interests.* Toronto, Ontario: University of Toronto Press, 2001. Collection of ten papers presented at a conference in Ottawa in late 1999, offering a range of views.

Economides, Michael, and Ronald Oligney. *The Color of Oil: The History, the Money and the Politics of the World's Biggest Business.* Katy, Tex.: Round Oak Publishing Co., 2000. Describes the culture of the global oil industry, including the chief petroleum-producing countries and the multinational companies that seek access to oil around the world. The book claims that petroleum will remain the world's dominant fuel for many decades to come.

Epstein, Lita, C. D. Jaco, and Julianne C. Iwersen-Neimann. *The Complete Idiot's Guide to the Politics of Oil.* New York: Penguin Group/Alpha Books, 2003. Lively book for the general reader describes the history of the oil industry and its current and possible future place in U.S. and world politics.

Gold, Thomas. *The Deep Hot Biosphere: The Myth of Fossil Fuels.* New York: Copernicus Books, 2001. Maverick proposal that hydrocarbons are of geological rather than biological origin and, therefore, that the world will not run short of them.

Goodman, Sidney J. *Asleep at the Geiger Counter: Nuclear Destruction of the Planet and How to Stop It.* Nevada City, Calif.: Blue Dolphin Publishing, 2002. The author, both an engineer and an environmental activist, exposes what he sees as the dangers and dishonesty of the nuclear industry.

Goodstein, David. *Out of Gas: The End of the Age of Oil.* New York: W. W. Norton, 2004. Explains why oil will be exhausted within 10 years and projects two future alternate-fuel scenarios, a bleak one featuring coal and greenhouse-effect heating and a cautiously optimistic one, in which natural gas and nuclear power provide a bridge to renewable fuels.

Grimston, Malcolm C., and Peter Beck. *Civil Nuclear Energy: Fuel of the Future or Relic of the Past?* London: Royal Institute of International Affairs, 2001. Illuminates important issues in the debate about nuclear power by presenting opposing viewpoints.

Henderson, Harry. *Nuclear Energy: A Reference Book.* Santa Barbara, Calif.: ABC Clio, 2000. An encyclopedia describing various aspects of the industry and related issues.

Hill, C. R., et al., eds. *Nuclear Energy: Promise or Peril?* River Edge, N.J.: World Scientific, 1999. Objectively describes the main issues involved in the development and use of nuclear energy. The book is written for the general reader.

Energy Supply

International Energy Agency. *Nuclear Power: Sustainability, Climate Change, and Competition.* Paris: Organisation of Economic Cooperation and Development, 1998. Concludes that nuclear energy has environmental and economic advantages but is unlikely to play a major role in world energy supply unless problems such as waste disposal are solved and public perception improves.

Jasper, James M. *Nuclear Politics: Energy and the State in the United States, Sweden, and France.* Princeton, N.J.: Princeton University Press, 2000. Claims that cultural and political factors caused the nuclear policies of these three nations, which were similar in the early 1970s, to diverge substantially in later years.

Kursunoglu, Behram N., Stephan L. Mintz, and Arnold Perlmutter, eds. *The Challenges to Nuclear Power in the Twenty-First Century.* New York: Plenum Press, 2000. Collection of papers by nuclear experts from many parts of the world, presented at the International Energy Forum in 1999.

Leventhal, Paul, Sharon Tanzer, and Steven Dolley, eds. *Nuclear Power and the Spread of Nuclear Weapons: Can We Have One Without the Other?* Dulles, Va.: Brassey's, 2002. Essays stress the connection between nuclear power and the proliferation of nuclear weapons.

Loucks, Robert Alden. *Shale Oil: Tapping the Treasure.* Philadelphia: Xlibris Corp., 2002. Short history of oil shale industry projects in the 1970s and 1980s by a participant, who claims that the industry is environmentally sound and should be further developed.

Makhijani, Arjun, Scott Saleska, and Mak Rev. *The Nuclear Power Deception.* New York: Apex Press, 1999. Reveals what authors call the "mythology" of nuclear power, including the belief that this energy source will be cheap and safe.

Mitchell, John, ed. *The New Economy of Oil: Impacts on Business, Geopolitics and Society.* London: Royal Institute of International Affairs, 2001. Considers issues that will affect and be affected by the oil and natural gas industries in the next 20 years, including supply, transportation, prices, security, and acceptability. The book concludes that acceptability will be a greater limiting factor than supply.

Montague, Gilbert Holland. *The Rise and Progress of the Standard Oil Company.* New York: Arno Press, 1973. Reprint of a book originally published in 1903 presents the history of John D. Rockefeller's famous (and infamous) late-19th-century behemoth, the ancestor of most of today's multinational oil companies. It offers reasons why the company was so successful and was able to form such beneficial relationships with the railroads.

Morris, Robert C. *The Environmental Case for Nuclear Power: Economic, Medical, and Political Considerations.* St. Paul, Minn.: Paragon House, 2000.

Decries antinuclear activism and contends that nuclear power is safe, better for the environment than fossil fuels, and more economical than alternative energy sources such as solar power.

Myers, Norman, and Jennifer Kent. *Perverse Subsidies: How Misused Tax Dollars Harm the Environment and the Economy.* Washington, D.C.: Island Press, 2001. Claims that the United States and other countries support fossil fuel and other environmentally damaging industries with hidden subsidies.

Odell, Peter. *Fossil Fuel Resources for the 21st Century.* Brentwood, Essex, U.K.: Multi-Science Publishing Co., Ltd., 2004. Claims that there is not and never has been a shortage of world oil supplies.

Taverne, Bernard G. *Petroleum, Industry and Governments: An Introduction to Petroleum Regulation, Economics and Government Policies.* New York: Kluwer Law International, 2000. Offers an overview of regulatory, economic, and policy forces that have shaped international oil and gas markets.

Yeomans, Matthew. *Oil: Anatomy of an Industry.* New York: New Press, 2004. Book critical of the industry includes history of oil and gasoline production, importance of oil in U.S. economy and politics, multinational oil corporations and their relationship with oil-producing countries, and reasons why continued dependence on oil is a liability.

Yergin, Daniel. *The Prize: The Epic Quest for Oil, Money and Power.* New York: Free Press (reissue), 1993. Comprehensive history of the oil industry.

ARTICLES

Adler, Kevin. "Oil Refining Changes Transportation, History, and Ways of Life." *Chemical Week,* vol. 164, September 18, 2002, pp. SS30–47. Extensive article on the history of the National Petroleum Refiners Association explains how the automobile and oil refining industries grew together and supported each other during the 20th century.

Appenzeller, Tim. "The End of Cheap Oil." *National Geographic,* vol. 205, June 2004, pp. 84–109. New technology may postpone the day, but easily extractable oil will run out—and probably sooner rather than later.

Barlett, Donald L., and James B. Steele. "The U.S. Is Running Out of Energy." *Time,* vol. 162, July 21, 2003, pp. 36 ff. Lack of a coherent federal energy policy leaves Americans hostage to dwindling oil and natural gas supplies.

Campbell, C. J. "The Supply Side: Before It's Too Late." *Newsweek International,* February 16, 2004, p. 42. Oil production is now reaching its peak and is about to begin declining.

"Canada: Oil Sands Project Hits Massive Cost Overrun." *Petroleum Economist,* vol. 71, April 2004, p. 38. Some parts of Alberta's massive tar (oil) sands project may cost almost twice the amount originally projected.

Cohn, Ernst M. "Energy Economics, 101." *Energy*, vol. 28, Summer 2003, pp. 45–47. Hydrocarbon fuels are not about to run out and are far more efficient sources of energy than hydrogen or other renewable fuels.

Creswell, Julie. "Oil Without End?" *Fortune*, vol. 147, February 17, 2003, p. 46. A few revisionist scientists say that petroleum is formed from geological rather than biological sources and is constantly being renewed.

Feltus, Anne. "Golden Oldies." *Petroleum Economist*, vol. 71, January 2004, pp. 16–17. Small, independent oil companies are rushing into areas around the world where the international giants do not bother to go, including parts of the Gulf of Mexico, the North Sea, West Africa, and Brazil.

Grossman, Karl. "The Nuclear Phoenix." *E*, vol. 12, November–December 2001, pp. 34–35. The George W. Bush administration is strongly pushing a revival of nuclear power, but environmentalists still distrust this energy source.

Guterl, Fred. "When Wells Go Dry." *Newsweek*, April 15, 2002, p. 32B. Using the methodology of famed geophysicist M. King Hubbert, Kenneth Deffeyes of Princeton University predicts that world oil production will peak between 2004 and 2008 and decline thereafter.

Hill, Patrice. "Energy Crisis Rekindles Interest in Nuclear Power." *Insight on the News*, vol. 17, April 23, 2001, pp. 31–32. Improved economics and safety of power plants may lead to new support for nuclear energy.

Hylton Hilary, et al. "Nuclear Summer." *Time*, vol. 157, May 28, 2001, p. 58. Improved safety and economics are making private industry and the federal government reconsider nuclear power.

Kriz, Margaret. "Showdown on the North Slope." *National Journal*, vol. 33, March 3, 2001, p. 634. Politicians are debating the possibility of drilling for oil under part of the Alaska National Wildlife Refuge.

Lake, James A., Amory Lovins, and Hunter Lovins. "Symposium." *Insight on the News*, vol. 17, August 27, 2001, pp. 40–41. Pro (Lake) and con (Lovins and Lovins) views on whether nuclear power is a good solution to energy shortages.

Lask, Ellen. "In Vogue, for Now." *Petroleum Economist*, vol. 69, January 2002, pp. 31–34. The outlook for the coal industry is good in the near term, but the effects of environmental regulations and competition from natural gas may eventually reduce coal's importance as an energy source.

Mann, Charles C. "Getting over Oil." *Technology Review*, vol. 105, January–February 2002, pp. 32–37. Cheap and abundant petroleum supplies are impeding the technical experimentation needed to develop renewable energy sources and curb global warming.

Maugeri, Leonard. "The Shell Game." *Newsweek International*, February 16, 2004, p. 40. Oil giant Shell's admission that it had substantially overstated its "proven reserves" should not be taken as evidence of either dishonesty or a coming shortage of oil.

"Nuclear Power: The Future?" *Power Economics*, vol. 5, April 2001, pp. 16–17. Nuclear power could help to combat electricity shortages and carbon dioxide emissions, but public concern about the economics and safety of this energy source will probably keep it from being employed.

Oliver, Mike, and John Hospers. "'Alternative Fuels'?" *The American Enterprise*, vol. 12, September 2001, pp. 20 ff. Nuclear power is a better choice for meeting the country's energy needs than alternative energy sources.

Parker, Harry W. "After Petroleum Is Gone, What Then?" *World Oil*, vol. 222, September 2001, pp. 70 ff. Analyzes seven organic carbon sources and processes for converting them to liquid fuels that might be used in transportation as substitutes for gasoline.

Rauber, Paul. "Snake Oil for Fossil Fools." *Sierra*, vol. 86, May 2001, pp. 56–57. Holds that fossil fuel industries, not environmentalists, are responsible for oil and electricity shortages.

Rauch, Jonathan. "The New Old Economy: Oil, Computers, and the Reinvention of the Earth." *Atlantic Monthly*, vol. 287, January 2001, pp. 35–49. Tells how new, computer-based techniques for oil exploration and drilling are greatly increasing the industry's productivity and the prospects for future oil supplies.

Savoye, Bruno, et al. "Exploring the Ocean's Depths." *Petroleum Economist*, vol. 70, April 2003, pp. 18–19. Oil companies are exploring the deep sea, especially off the coast of West Africa, and finding promising new petroleum deposits.

Shook, Barbara. "Spindletop Changed World 100 Years Ago Today." *Oil Daily*, vol. 51, January 10, 2001, n.p. The explosive birth of this giant oil field near Beaumont, Texas, on January 10, 1901, is said to mark the beginning of the modern American petroleum industry.

"Strategic Deterrent." *Global Markets*, vol. 33, May 5, 2003, p. 1. There is confusion about the real purpose of the U.S. Strategic Oil Reserve.

Thomas, Victoria. "Heading for Compromise." *Petroleum Economist*, vol. 68, July 2001, p. 12. Analysis of the federal National Energy Policy Development Group's proposed national energy policy focuses on ways to increase the supply of domestic energy sources, especially natural gas.

Toal, Brian A. "The Oil and Gas Game." *Oil and Gas Investor*, vol. 21, January 2001, pp. 76 ff. A chronology of developments in the oil and natural gas industries between 1980 and 2001.

Tsubata, Kate. "Nuclear Power's New Promise and Peril." *World and I*, vol. 17, May 2002, n.p. The dangers of nuclear power may have been exaggerated, but the industry still faces serious safety issues, including the possibility of a terrorist attack.

West, J. Robinson. "Five Myths About the Oil Industry." *The International Economy*, vol. 17, Summer 2003, pp. 45–47. Claims that five common negative beliefs about the oil industry are not correct.

Williams, Peggy. "Good Well Hunting." *Oil and Gas Investor*, vol. 23, December 2003, pp. 37–40. New technology and determined exploration are constantly uncovering new oil and gas deposits, especially in the deep sea and the Arctic, but extracting this bounty will be neither easy nor inexpensive.

"Will the Oil Run Out?" *The Economist*, February 10, 2001, p. 5. Doomsayers are wrong in predicting that oil will either run out or be replaced as the world's chief fuel in the next 50 years, although people may begin to turn away from hydrocarbons for reason other than scarcity.

Web Documents

American Institute of Petroleum. "Offshore Drilling." World Petroleum Congress web site. Available online. URL: http://www.world-petroleum.org/education/offdrill/index.html. Downloaded on July 23, 2004. Describes how offshore drilling takes place, environmental safeguards, drilling rig types, production platforms, and waste products.

———. "Refining of Petroleum." World Petroleum Congress web site. Available online. URL: http://www.world-petroleum.org/education/petref/index.html. Downloaded on July 23, 2004. Describes refining processes, products of refining, and refineries and the environment.

Elliot, R. Neal, et al. "Natural Gas Price Effects of Energy Efficiency and Renewable Energy Practices and Policies." American Council for an Energy-Efficient Economy. Available online. URL: http://www.aceee.org/pubs/e032full.pdf. Posted in December 2003. Examines price effects of efficiency and renewable energy programs in different states. The site shows that energy efficiency and renewable energy can reduce natural gas prices and price volatility.

"Energy Supply Prospects and Politics: Focus on Alaska." Petroleum Industry Research Foundation, Inc. Available online. URL: http://www.pirinc.org/download/alaskaoilandgasapril03.pdf. Posted in April 2003. Discusses issues involved in bringing additional oil and gas from Alaska's North Slope, including the Alaska National Wildlife Refuge, to market and the need to do so in light of unrest or war in Venezuela, Iraq, Nigeria, and other sources of imported oil.

"Environmental Benefits of Advanced Oil and Gas Exploration and Production Technology." U.S. Department of Energy, Office of Fossil Energy. Available online. URL: http://www.fossil.energy.gov/programs/oilgas/publications/environ_benefits/Environmental_Benefits_Report.

html. Posted in October 1999. Highlights 36 technological innovations as samples of improvements that have increased the industry's efficiency and decreased its damage to the environment during the last 30 years.

Ferguson, Rich. "Risky Diet 2003: Natural Gas: The Next Energy Crisis." Center for Energy Efficiency and Renewable Technologies. Available online. URL: http://www.ceert.org/pubs/crrp/natgas/riskydiet2003.pdf. Posted in September 2003. Describes forces likely to drive up demand for natural gas. Ferguson claims that recent gas shortages and high prices are permanent and could lead to another energy crisis like the one that struck electricity in California in 2000–01.

———. "Risky Diet: North America's Growing Appetite for Natural Gas." Center for Energy Efficiency and Renewable Technologies. Available online. URL: http://www.ceert.org/pubs/crrp/natgas/riskydietreport.pdf. Posted in April 2002. Describes economic and environmental problems with currently growing demand for natural gas and urges use of renewable energy sources instead.

"Fusion Energy (Position Statement 12)." American Nuclear Society. Available online. URL: http://www.ans.org/pi/ps/docs/ps12.pdf. Last updated November 2003. Provides background on the fusion process and a description of progress in fusion research.

"Guide to Nuclear Energy." Nuclear Energy Institute. Available online. URL: http://www.nei.org/documents/guidetonuclearenergy.pdf. Posted in January 2001. Includes a vision of a nuclear energy future, plant safety, the next generation of nuclear power plants, regulation, managing and transporting used nuclear fuel, low-level radioactive waste management, and the economic value of emission-free nuclear power.

Lochbaum, David. "U.S. Nuclear Plants in the 21st Century: The Risk of a Lifetime." Union of Concerned Scientists. Available online. URL: http://www.ucsusa.org/documents/nuclear04fnl.pdf. Posted in May 2004. The risks for catastrophe change as nuclear reactors age, so plants must be aggressively monitored during all three stages of their lifetime: the break-in phase, middle-life phase, and wear-out phase. Lochbaum identifies the best ways to manage risks at different stages.

"The Long Term Storage of Radioactive Waste: Safety and Sustainability." International Atomic Energy Agency. Available online. URL: http://www-pub.iaea.org/MTCD/publications/PDF/LTS-RW_web.pdf. Posted in June 2003. International experts discuss types of facilities and factors influencing decisions about which type to use.

MacKenzie, James J. "Oil as a Finite Resource." World Resources Institute. Available online. URL: http://www.wri.org/pubs_content.cfm?PubID=2688. Posted in March 2000. Concludes that world oil production will peak between 2007 and 2013.

"New Technology." American Petroleum Institute. Available online. URL: http://api-ec.api.org/policy/index.cfm?objectid=FC39AA1D-7CA7-11D5-BC6B00B0D0E15BFC&method=display_body&er=1&bitmask=001001001000000000. Downloaded on July 22, 2004. Lists recent technological advances in petroleum exploration, production, refining, and transportation.

"Nuclear Energy in a Sustainable Development Perspective." Nuclear Energy Agency of the Organisation for Economic Cooperation and Development. Available online. URL: http://www.nea.fr/html/ndd/docs/2000/nddsustdev.pdf. Posted in 2000. Report intended to help governments decide whether, and to what degree, nuclear power is compatible with goals of sustainable development; discusses this energy source from economic, environmental, and social viewpoints.

"Nuclear Technology Review:1 2003 Update." International Atomic Energy Agency. Available online. URL: http://www.iaea.or.at/OurWork/ST/NE/Pess/assets/ntr2003.pdf. Posted in September 2003. This review, issued every two years and updated annually, reports on the global status and trends in fields of nuclear science and technology. Topics covered include new technology; applications in health, agriculture, water, and other areas; information and knowledge management; and the role of nuclear power in sustainable development.

"The Pebble Bed Modular Reactor (PBMR)." Nuclear Information and Resource Service. Available online. URL: http://www.nirs.org/factsheets/PBMRFactSheet.htm. Downloaded on July 23, 2004. Claims that the pebble-bed modular reactor, hailed by the nuclear industry as a newer and safer type of nuclear power plant, is merely "old wine in new bottles" and is still fundamentally unsafe.

"Policy Direction for Canada's Oil and Gas Industry." Canadian Association of Petroleum Producers. Available online. URL: http://www.capp.ca/raw.asp?x=1&dt=NTV&e=PDF&dn=67582. Posted in September 2003. Trade industry's submission to the Council of Energy Ministers discusses Canada's oil and gas resources and their place in the Canadian and world economies, as well as making recommendation for national policy.

"Questions and Answers About Natural Gas and National Energy Policy." American Gas Association. Available online. URL: http://www.aga.org/Content/ContentGroups/Home_Page/Limelight/Questions_and_Answers_About_Natural_Gas_and_National_Energy_Policy.htm. Downloaded on July 22, 2004. Provides background on extraction, use, and benefits of natural gas and lists this gas trade association's recommendations for a national energy policy.

"Timeline: The Nuclear Waste Policy Dilemma." Eureka County Nuclear Waste. Available online. URL: http://www.yuccamountain.org/time.htm.

Downloaded on April 15, 2004. Timeline of events related to a congressional decision to set up a permanent storage facility for highly radioactive waste from nuclear power plants at Yucca Mountain, Nevada.

"Vision 2020: Powering Tomorrow with Clean Nuclear Energy." Nuclear Energy Institute. Available online. URL: http://www.nei.org/index.asp?catnum=2&catid=143. Downloaded on July 23, 2004. Describes and provides background on the nuclear energy industry's Vision 2020 program, stressing the advantages of nuclear power as an electricity source that does not emit greenhouse gases.

"A Young Person's Guide to Oil and Gas." World Petroleum Congress. Available online. URL: http://www.world-petroleum.org/education/ip1/ip1.html. Downloaded on March 31, 2004. Illustrated with many drawings and diagrams, this brief guide explains the importance of oil and gas; their origin; how they are recovered from the ground, transported, and refined; what products are made from them; and how these products reach customers.

RENEWABLE ENERGY SOURCES AND ENERGY CONSERVATION

BOOKS

Asmus, Peter. *Reaping the Wind: How Mechanical Wizards, Visionaries, and Profiteers Helped Shape Our Energy Future.* Washington, D.C.: Island Press, 2000. History of the commercial wind power industry in the United States, including the individuals, technologies, and policies that have driven it.

Augustyn, Jim. *Return of the Solar Cat Book: Mixing Cat Wisdom with Science and Solar Politics.* Revised edition. Berkeley, Calif.: Patty Paw Press, 2003. Uses a humorous approach featuring cats to explain the nature and advantages of solar energy in particular and renewable energy and conservation in general.

Berinstein, Paula. *Alternative Energy: Facts, Statistics, and Issues.* Phoenix: Oryx Press, 2001. Focuses on alternative energy sources but also analyzes conventional sources. Berinstein provides numerous statistical tables as well as succinct descriptions and evaluations of technological, economic, and environmental features of each source.

Braun, Harry. *The Phoenix Project: Shifting from Oil to Hydrogen.* Revised edition. Phoenix: SPI Publications, 2000. Urges building of massive infrastructure to transport and use hydrogen, which will be obtained from water by wind power.

Cassedy, Edward S. *Prospects for Sustainable Energy: A Critical Assessment.* New York: Cambridge University Press, 2000. Book aimed at the general reader emphasizes the need to use alternative fuel sources and evaluates them in terms of sustainability, describing the origins, technology, marketability, and environmental impacts of each type.

Casten, Thomas R. *Turning Off the Heat: Why America Must Double Energy Efficiency to Save Money and Reduce Global Warming.* Loughton, Essex, U.K.: Prometheus Books, 1998. Energy conversion entrepreneur shows how his company lowered costs and reduced pollution through combined heat and power generation. He proposes a regulatory method to increase overall energy efficiency.

Convery, Frank J., ed. *A Guide to Policies for Energy Conservation: The European Experience.* Northampton, Mass.: Edward Elgar, 1999. Analysis of European experiences shows that the design and execution of policies to promote energy conservation are more important than the policies' content in determining their success.

Cothran, Helen, ed. *Energy Alternatives: Opposing Viewpoints.* San Diego: Greenhaven Press, 2002. Pro and con articles on the necessity of using alternative energy sources, the viability of nuclear energy, the value of different alternative sources, and the usefulness of alternative vehicle fuels.

Dunn, Seth. *Hydrogen Futures: Toward a Sustainable Energy System.* Washington, D.C.: Worldwatch Insttitute, 2001. Prefers direct use of hydrogen as a vehicle fuel to making the gas onboard vehicles from gasoline or methanol through reformers. Dunn stresses the importance of government support in developing a hydrogen economy.

Edwards, Brian K. *The Economics of Hydroelectric Power.* Northampton, Mass.: Edward Elgar, 2003. Discusses the role of dams, including their environmental effects, and the place of hydropower in the electricity industry and its regulation. Edwards includes a dynamic model of a hydroelectric generating facility and case studies of dams in the United States.

European Association of Renewable Energy Research Centers. *The Future of Renewable Energy 2.* Brussels, Belgium: European Association of Renewable Energy Research Centers, 2004. Examines each of the major renewable energy sources in detail and lays out a path to more widespread use of renewable technologies.

Ewing, Rex A. *Power with Nature: Solar and Wind Energy Demystified.* LaSalle, Colo.: Pixyjack Press, 2003. Practical information on these two renewable energy sources by someone who obtains all his electricity from them.

Froschauer, Karl. *White Gold: Hydroelectric Power in Canada.* Vancouver: University of British Columbia, 1999. Applies political and economic perspectives to analysis of five large hydroelectric projects to show what went wrong with hydropower development in Canada. Froschauer claims that integrating provincial projects into a national network would have greatly improved the situation.

Annotated Bibliography

Geller, Howard. *Energy Revolution: Policies for a Sustainable Future.* Washington, D.C.: Island Press, 2002. Analyzes policies and programs used in different countries, especially the United States and Brazil, to encourage a sustainable pattern of energy use based on efficiency and renewable sources.

Gordon, Jeffrey M., ed. *Solar Energy—The State of the Art.* Freiburg, Germany: International Solar Energy Society, 2001. Contains chapters devoted to 12 different solar energy subdisciplines, focusing on technology.

Hayden, Howard C. *The Solar Fraud: Why Solar Energy Won't Run the World.* Pueblo West, Colo.: Vales Lake Publishing, 2002. Claims that laws of physics explain why solar and wind power do not, and never will, contribute significantly to the world's energy supply.

Hoffman, Peter. *Tomorrow's Energy: Hydrogen, Fuel Cells, and the Prospects for a Cleaner Planet.* Cambridge, Mass.: MIT Press, 2001. Describes the advantages of a hydrogen-based economy (especially when the hydrogen is obtained from water with solar or other renewable energy) and progress toward reaching that goal.

Hordeski, Michael F. *New Technologies for Energy Efficiency.* New York: Marcel Dekker, 2002. Engineer examines new technologies for increasing energy efficiency cost effectively and reducing dependence on the power grid.

Laird, Frank N. *Solar Energy, Technology Policy, and Institutional Values.* New York: Cambridge University Press, 2001. Analyzes renewable energy policy in the United States from the end of World War II to the end of the 1970s.

Malkina-Pykh, Irina G., and Yuri A. Pykh. *Sustainable Energy: Resources, Technology and Planning.* Ashurst, Southampton, U.K.: WIT Press, 2002. Discusses fundamental considerations associated with energy sustainability and evaluates stationary (nonvehicular) uses of renewable energy sources in different parts of the world.

National Research Council. *Hydrogen Economy: Opportunities, Costs, Barriers, and R&D Needs.* Washington, D.C.: National Academies Press, 2004. Study suggests that hydrogen is a potentially beneficial long-term approach to energy, but transition to a hydrogen economy will take a long time. Meanwhile, the U.S. government should encourage research into a number of alternative energy sources and methods of increasing energy efficiency.

Organisation for Economic Cooperation and Development. *Creating Markets for Energy Technologies.* Paris: Organisation for Economic Cooperation and Development, 2003. Papers from a 2001 workshop present and evaluate programs to expand markets for efficient and environmentally benign energy technologies.

Owen, Gill. *Public Purpose or Private Benefit? The Politics of Energy Conservation.* Manchester, U.K.: Manchester University Press, 1999. Examines politics and policies of energy conservation during the last 30 years in the Netherlands, Denmark, the United States, Britain, Japan, and Australia.

Parmon, V. N., et al., eds. *Chemistry for the Energy Future.* Oxford, U.K.: Blackwell Science, 2000. Stresses chemistry's contribution to adoption of renewable energy sources but also examines the role of chemistry in nuclear energy and more efficient combustion of fossil fuels.

Perlin, John. *From Space to Earth: The Story of Solar Electricity.* Cambridge, Mass.: Harvard University Press, 2002 (reprint). Provides a history of photovoltaics (solar cells), emphasizing their successes.

Righter, Robert W. *Wind Energy in America: A History.* Norman: University of Oklahoma Press, 2003. Considers both technical and political influences. Righter focuses on wind energy as a way of providing decentralized power to the rural United States.

Romm, Joseph J. *The Hype About Hydrogen: Fact and Fiction in the Race to Save the Climate.* Washington, D.C.: Island Press, 2004. Expresses pessimism about the usefulness of hydrogen as an automobile fuel, but holds out hope that it will be a useful source of electricity.

Sayigh, A. A. M., ed. *Renewable Energy: The Energy for the 21st Century.* New York: Pergamon Press, 2000. Collection of papers from the sixth World Renewable Energy Congress, held in July 2000, focuses on climate change as a spur to promotion of renewable energy sources.

Scheer, Hermann. *The Solar Economy: Renewable Energy for a Sustainable Global Future.* London: Earthscan Publications, 2002. Originally published in German in 1999, this book by a member of the German parliament stresses the importance of renewable, especially solar, energy and offers an alternative to the Kyoto Protocol.

Sims, Ralph. *Bioenergy Options for a Cleaner Environment: In Developed and Developing Countries.* St. Louis, Mo.: Elsevier Science, 2003. Panel of international experts describes range of present and future biomass technologies, as well as presenting barriers and advantages to using this energy source in developed and developing countries.

Transportation Research Board. *Effectiveness and Impact of Corporate Average Fuel Economy (CAFE) Standards.* Washington, D.C.: National Academies Press, 2002. National Academy of Science and U.S. Department of Transportation study of the impact of regulation of automotive fuel economy standards on vehicle safety, the economy of the automotive sector and employment in that sector, and the auto consumer.

Walisiewicz, Marek. *Alternative Energy: A Beginner's Guide to the Future of Energy Technology.* New York: DK Publishing, 2002. Brief but well-illustrated and fact-packed book introduces alternative energy sources and compares them to conventional ones.

Wilkins, Gill. *Technology Transfer for Renewable Energy: Overcoming Barriers in Developing Countries.* London: Earthscan Publications, 2002. Explores role of renewable energy, especially solar and biomass, in sustainable development, focusing on rural areas of developing countries. Wilkins includes case studies and analysis of the Kyoto Protocol's Clean Development Mechanism.

Winebrake, James J., ed. *Alternate Energy: Assessment and Implementation Reference Book.* Lilburn, Ga.: Fairmont Press, 2003. Thoroughly assesses the technology of different alternate energy sources and its implementation. The book is international in scope and cautiously optimistic.

ARTICLES

Adrian, Clare. "Power Boost." *Recycling Today,* vol. 42, May 2004, pp. 94–98. A need for new energy sources could revive interest in waste-to-energy plants. They are cleaner than they used to be, but they are still expensive.

Anderson, Bruce. "The Energy for Battle." *Spectator,* vol. 287, December 8, 2001, pp. 20–21. A layperson's explanation of the nature, advantages, and challenges of the hydrogen fuel cell and the new society it could create.

Anderson, Heidi. "Environmental Drawbacks of Renewable Energy." *Energy User News,* vol. 26, June 2001, pp. 30 ff. Environmental drawbacks of renewable energy sources are real, but they are insignificant compared to the drawbacks of conventional sources.

Ayer, Fred, and Lydia Grimm. "Promoting a Clean, Green, Image." *International Water Power and Dam Construction,* vol. 54, March 2002, pp. 28–30. The Low Impact Hydropower Institute has created a voluntary certification program that would allow qualifying facilities to market themselves as "green."

Ayres, Robert U. "The Energy We Overlook." *World Watch,* vol. 14, November–December 2001, pp. 30–39. Considers conservation, alternative fuels, and sequestration as ways of reducing carbon dioxide emissions and concludes that conservation works best.

Bird, Maryann, et al. "Selling the Sun . . . and the Wind." *Time,* vol. 158, July 16, 2001, pp. B8–9. Renewable energy has come of age, but only foreign companies are likely to make money on it unless the U.S. government supports it, or at least stops discriminating against it.

Bivens, Matt. "Fighting for America's Energy Independence." *The Nation,* vol. 274, April 15, 2002, pp. 11–12. Technology has greatly improved the economic viability of renewable energy sources, but the

U.S. government needs to at least stop subsidizing fossil fuels, and preferably increase subsidies and incentives for renewables, in order to make the change to renewables happen.

Black, Jane. "Why Wait for Hydrogen to Kick In?" *Business Week*, September 9, 2003, n.p. Hydrogen-powered cars will not be practical for a decade or more, but technologies exist that could make cars more fuel efficient today.

Blundell, Tom. "Action Must Start Now." *New Statesman*, vol. 132, February 24, 2003, pp. xviii–xix. The Royal Commission on Environmental Pollution concludes that both increased use of noncarbon energy sources and energy conservation will be necessary if Britain is to do its part to reduce carbon dioxide emissions.

"Britannia Rules the Waves." *Power Engineering International*, vol. 12, May 2004, pp. 57–59. British companies are leaders in development of wave and tidal energy technologies, but they desperately need funding to continue their progress.

Caldwell, Jim. "Rising Wind—Time to Take a Closer Look." *Power Engineering*, vol. 107, September 2003, p. 69. Economic and technical problems caused by wind's variability are not as great as critics have claimed.

Cassidy, Peter, and Mark Raymont. "The Effects of Energy Policy on the Future of Hydro-Electricity." *Power Economics*, vol. 5, February 2001, pp. 26–27. Critics' claims that hydropower may add to the greenhouse effect, as well as new and proposed European legislation, may limit the use of this technology.

Collins, Lyn. "Renewable Energy: What Are the Options?" *Geography Review*, vol. 15, November 2001, pp. 20–23. Examines the scope for development of various renewable energy sources in Britain.

———. "Renewable Energy: Why Should We Use It?" *Geography Review*, vol. 15, September 2001, pp. 2 ff. Considers environmental, political, and economic reasons for using renewable energy.

Dauncey, Guy. "A Sustainable Energy Plan for the U.S." *Earth Island Journal*, vol. 18, Autumn 2003, pp. 32–35. The United States can meet its energy needs through renewable resources if government energy policy favors such sources.

"Energy: No Renewable Energy Targets Beyond 2010." *European Report*, May 8, 2004, p. 403. Summarizes a recent European Commission report on progress toward renewable energy goals, which states that the goals are attainable but only with considerable effort.

Fetters, John. "Fuel Cells: Right Out of Jules Verne." *Energy User News*, vol. 27, September 2002, p. 10. Succinct review of the nature and applications of hydrogen fuel cells and challenges to their widespread use.

Annotated Bibliography

Ford, Neil. "The Need for Alternatives to HEP." *African Business*, August–September 2003, pp. 36–37. Africa has huge hydroelectric power (HEP) capacity, but it should diversify its energy sources by developing small plants that use natural gas or renewable sources such as wind power.

"Fuel Economy: Stalled in Traffic." *Consumer Reports*, vol. 67, December 2002, pp. 56–59. Steps can and should be taken to increase the fuel economy of cars, light trucks, and SUVs.

Galiano, Troy. "Renewing the Energy Debate." *State Legislatures*, vol. 28, April 2002, pp. 14–19. Many states are expanding their use of renewable energy sources because doing so can protect them against surges in energy costs and also generate revenue.

"Green Nations." *Geographical*, vol. 75, August 2003, pp. 82–83. Reviews various nations' successes in using renewable energy sources, which are usually spurred by the high cost of conventional energy and by government and public support.

Grose, Thomas K. "Wind Turbines, Solar Panels and Fuel Cells—Is Alternative Energy Ready for Prime Time?" *Time International*, vol. 158, December 31, 2001, p. 108. Renewable energy sources face economic problems, but these are solvable, and it is in the interest of national security to do so.

Hirsch, Robert L. "Large-Scale Green Power: An Impossible Dream?" *Public Utilities Fortnightly*, vol. 141, January 1, 2003, pp. 25–27. It is highly unlikely that wind and solar cells will ever provide a cost-effective alternative to fossil fuels as a major source of electricity generation because of the unreliability of these energy sources.

Hopkins, Barry, Carolyn Orr, and Scott Richards. "Renewable Energy Gains Steam." *State Government News*, vol. 46, May 2003, pp. 29–30. Describes environmental, security, and economic benefits of renewable energy sources and lists several types of incentives that states in the United States are applying to encourage use of these sources.

Huber, Peter. "The Efficiency Paradox." *Forbes*, vol. 168, August 20, 2001, p. 64. Paradoxically, increased efficiency leads to greater total energy use because it makes energy-consuming activities cheaper.

Jonas, Junona. "Primer on Geothermal Energy." *Electric Light & Power*, vol. 81, November 2003, p. 30. Basic information about geothermal energy and its use in the United States today.

Kleit, Andrew N. "Impacts of Long-Range Increases in the Fuel Economy (CAFE) Standard." *Economic Inquiry*, vol. 42, April 2004, pp. 279–294. Maintains that a stiff gasoline tax would save just as much gasoline as increasing the fuel economy standard and cause less damage to the economy.

Labs, Wayne. "Fuel Cells Go Back to the Future—Again." *Electronic Design*, vol. 52, June 14, 2004, pp. 83–85. Hydrogen fuel cells and solar photovoltaic cells can reduce consumption of nonrenewable energy sources, but they need help from both government and private sectors.

Laurent, Christine. "Beating Global Warming with Nuclear Power?" *UNESCO Courier*, February 2001, pp. 37–38. Greater use of nuclear power could lead to a reduction in greenhouse gas emissions, but increasing energy efficiency may be a better way to accomplish the same end.

Lewis, Dan. "Blooming Awful?" *Power Economics*, vol. 8, April 2004, pp. 13–14. Criticizes Britain's government for relying on taxing and regulation rather than the market to promote renewable energy.

Lovins, Amory. "Old Problems, New Solutions." *WorldLink*, vol. 15, July–August 2002, pp. 42–47. Increased energy efficiency can allow economic growth to coexist with reduction in energy use.

Lovins, Amory, and L. Hunter Lovins. "Energy Forever." *The American Prospect*, vol. 13, February 11, 2002, pp. 30–34. A combination of increased energy efficiency and decentralized electricity generation can save money, help the environment, and reduce the vulnerability of the energy supply to terrorism.

Madsen, Birger T. "Energy's Winds of Change." *UNESCO Courier*, March 2000, pp. 9–10. A wind power supporter describes the industry's rapid advances worldwide and its hopes for the future.

Mandelbaum, Robb. "Greenmark." *Discover*, vol. 25, June 2004, pp. 48–55. Describes Denmark's experiment with wind, biomass, and other renewable energy resources on the island of Samsø.

McKibben, Bill. "It's Easy Being Green." *Mother Jones*, vol. 27, July–August 2002, pp. 34–38. Americans could save energy and cut fossil fuel use by many easy steps that would not substantially change lifestyles—and some are doing so, but the George W. Bush administration is not among them.

McVicar, Euan. "Unsure Wind." *Utility Week*, June 25, 2004, p. 18. Claims that offshore wind energy projects in Britain are promising but need more certainty of government support in order to develop.

Ottinger, Richard L., and Rebecca Williams. "Renewable Energy Sources for Development." *Environmental Law*, vol. 32, Spring 2002, pp. 331–368. Renewable energy sources hold particular promise in developing countries, but many economic, legal, and social barriers limit their adoption. Authors explore legal mechanisms for overcoming these barriers and provide examples of countries that have used them.

Parker, Paul, Ian S. Rowlands, and Daniel Scott. "Innovations to Reduce Residential Energy Use and Carbon Emissions: An Integrated Approach." *The Canadian Geographer*, vol. 47, Summer 2003, pp. 169–184.

Discusses a Canadian program to increase energy efficiency of homes and considers reasons for its success, which include participation of multiple local stakeholders and consideration of social as well as technical factors that determine whether homeowners will implement recommended changes.

Rigden, Dan. "The Green Challenge." *Petroleum Economist*, vol. 69, August 2002, p. 27. Renewable energy sources, especially wind, are likely to become more cost competitive with fossil fuels in the next 20 years, but they will still probably contribute only a small share of the global mix of energy sources.

———. "Majors Accept Green Fate." *Petroleum Economist*, vol. 69, August 2002, pp. 28–29. Many oil companies are investing in renewable fuels, but the degree and focus of their investment vary.

Roosevelt, Margot. "The Winds of Change." *Time*, vol. 160, August 26, 2002, p. A40. Wind is the world's fastest growing power source, but incorporating it and other renewables into the energy mix will require government action.

Schimmoller, Brian K. "Renewable Energy Enters Commercial Era." *Power Engineering*, vol. 108, April 2004, pp. 38–41. Commercially viable renewable energy is finally here. It will probably evolve in ways different from those that shaped the conventional energy sector.

Schmidt, Kira. "Fuel-Economy Falsehoods." *Earth Island Journal*, vol. 17, Spring 2002, p. 8. Measurements of cars' expected fuel economy are inaccurate because they are based on driving conditions that no longer exist.

Sheppard, Robert. "Beyond Kyoto." *Maclean's*, November 11, 2002, pp. 18–19. Canadians could help their country meet its greenhouse gas reduction goals by many small conservation-inducing changes in their homes and cars—but will they?

Skinner, Claire. "When the Wind Grows." *Utility Week*, August 8, 2003, p. 26. Renewable energy use is slated to increase greatly in Europe, especially wind power in Britain and biomass on the continent, creating many jobs.

Stipp, David. "Can This Man Solve America's Energy Crisis?" *Fortune*, vol. 145, May 13, 2002, pp. 100–101. Amory Lovins thinks that energy efficiency is the best way to save money and the environment at the same time—and many businesses have come to agree with him.

Taylor, Jerry, and Peter VanDoren. "Evaluating the Case for Renewable Energy." *Power Economics*, vol. 6, March 2002, pp. 24–27. The high cost of renewable energy sources keeps them from being competitive with fossil fuels now and probably will continue to do so in the future.

"The Technologies." *Geographical*, vol. 75, August 2003, pp. 78–79. Surveys present and future technology and uses of solar, wind, and biomass power.

Underhill, William. "Taking the Breeze." *Newsweek,* April 15, 2002, p. 32. Improved turbine designs and siting make wind farms more efficient, but aesthetics and economics can still work against them.

Wee, Heesun. "Can Oil Giants and Green Energy Mix?" *Business Week Online,* September 25, 2002, n.p. Some multinational oil companies are making substantial investments in renewable energy sources so that the companies can continue to make a profit even if these sources begin to replace oil.

WEB DOCUMENTS

Aitken, Donald W. "Transitioning to a Renewable Energy Future." International Solar Energy Society. Available online. URL: http://www.ises.org/shortcut.nsf/to/wp. Posted in 2003. Discusses elements driving public policy toward a renewable energy transition; the characteristics, status of development, and potential of various types of renewable energy; national and local factors supporting the development and application of renewable energy technologies; policies and market incentives to accelerate the application of renewable energy resources; the role of research and development; and the United States and Germany as models of national clean energy policy.

"AWEA Wind Web Tutorial." American Wind Energy Association. Available online. URL: http://www.awea.org/faq/index.html. Posted in 2004. Includes wind energy basics, costs, potential, wind energy and the economy, wind energy and the environment, offshore wind, wind industry statistics, small wind energy systems, policy issues, and a resource guide.

Bauen, Ausilio, Jeremy Woods, and Rebecca Hailes. "Biopowerswitch! A Biomass Blueprint to Meet 15% of OECD Electricity Demand by 2020." World Wildlife Organization. Available online. URL: http://www.panda.org/downloads/europe/biomassreportfinal.pdf. Posted in April 2004. Defines biomass, describes biomass-produced electricity in OECD countries and processes for producing electricity from biomass, compares agricultural residues and biomass-dedicated crops, lists benefits of using biomass as a fuel and of the specific 15 percent target, discusses economic and environmental effects of bioelectricity, and predicts future directions for the technology.

"Biomass as a Renewable Energy Source." Royal Commission on Environmental Pollution. Available online. URL: http://www.rcep.org.uk/biomass/Biomass%20Report.pdf. Downloaded on July 23, 2004. Makes recommendations for using biomass fuels in electricity generation.

"BlueAge: Blue Energy for a Green Europe." European Small Hydropower Association. Available online. URL: http://www.esha.be/BlueAge.pdf.

Downloaded on July 22, 2004. Strategic study for the development of small hydropower in the European Union.

Clemmer, Steven, et al. "Clean Energy Blueprint: A Smarter National Energy Policy for Today and the Future." Union of Concerned Scientists. Available online. URL: http://www.ucsusa.org/documents/blueprint.pdf. Posted in October 2001. Shows how consumers could save more than $440 billion between 2002 and 2020 if legislators enacted recommended energy-efficiency and renewable energy policies.

"Energy Security: Solutions to Protect America's Power Supply and Reduce Oil Dependence." Union of Concerned Scientists. Available online. URL: http://www.ucsusa.org/documents/acfwuam30.pdf. Posted in January 2002. Recommends government support for renewable energy sources because they will reduce oil dependence and infrastructure vulnerability, thus increasing energy security.

Gawell, Karl, Marshall Reed, and P. Michael Wright. "Preliminary Report: Geothermal Energy, the Potential for Clean Power from the Earth." Geothermal Energy Association. Available online. URL: http://www.geo-energy. org/PotentialReport.htm. Posted on April 7, 1999. Country-by-country estimates of geothermal energy's potential to meet electricity demand.

Gray, Thomas O. "Wind Energy: Views on the Environment: Clean and Green." American Wind Energy Association. URL: http://www.awea.org/pubs/documents/oppoll.PDF. Downloaded on July 22, 2004. Eleven-page summary of public opinion surveys on the environment in general, renewable energy in general, and wind energy in particular, using data gathered from polling in the United States, Britain, and Canada.

Gsaenger, Stefan. "Status and Perspective of the Wind Industry: An International Overview." World Wind Energy Association. Available online. URL: http://www.wwindea.org/default.htm. Posted in June 2003. Wind energy facts, including policies and economy, promotion, different types of systems and their efficiency, experiences, and statistical data.

"History: From the Water Wheel to Hydropower." Canadian Hydropower Association. Available online. URL: http://www.canhydropower.org/hydro_e/p_ hyd_a.htm. Downloaded on July 22, 2004. Brief history/chronology of hydroelectric power development in Canada.

"Hydropower: Another Look at a Natural Energy Resource for the Future." National Hydropower Association. Available online. URL: http://www.hydro.org/pdf/HydroPositioningPaper.pdf. Downloaded on July 23, 2004. Claims that hydropower has many advantages as an energy source and that accusations that it harms fish and other river life are exaggerated.

"Issues for Renewable Fuels in Competitive Electricity Markets." Energy Information Administration (Department of Energy). Available online. URL:

http://www.eia.doe.gov/cneaf/electricity/chg_str_fuel/html/chapter5.html. Last updated May 27, 2003. States that renewable energy sources are less harmful to the environment but more expensive than conventional ones, so government support for renewables is needed. The site surveys different types of support programs.

Lovins, Amory B. "Designing a Sustainable Energy Future—Integrating Negawatts with Diverse Supplies at Least Cost." Rocky Mountain Institute. Available online. URL: http://www.rmi.org/images/other/Energy/E03-13_DsnSusEnrgyFuture.pdf. Posted on November 13, 2003. Speech at Australia's first national conference on energy efficiency shows opportunities in advancing electric, heat, and transportation efficiency.

———. "Twenty Hydrogen Myths." Rocky Mountain Institute. Available online. URL: http://www.rmi.org/images/other/Energy/E03-05_20HydrogenMyths.pdf. Last updated September 2, 2003. Debunks popular misconceptions and claims that the path to a hydrogen economy is far more easy, attractive, and profitable than has been thought.

Palmer, Karen, and Dallas Burtraw. "Electricity, Renewables, and Climate Change: Searching for a Cost-Effective Policy." Resources for the Future. Available online. URL: http://www.rff.org/rff/Documents/RFF-RPT-Renewables.pdf. Posted in May 2004. Examines the effects of policy options designed to give renewables a larger share of the electricity market.

"Renewable Energy Scenario to 2040." European Renewable Energy Council. Available online. URL: http://www.erec-renewables.org/documents/targets_2040/EREC_Scenario%202040.pdf. Downloaded July 22, 2004. Predicts that half of the world's energy supply could come from renewable energy sources by 2040 and describes policy measures needed to achieve this challenging renewables goal.

"Scenarios for a Clean Energy Future." Oak Ridge National Laboratory. Available online. URL: http://www.ornl.gov/sci/eere/cef/. Posted in November 2000. Uses three scenarios, covering a 20-year time span, to evaluate costs and benefits of different policies to encourage energy efficiency and use of clean energy technologies.

Shibaki, Masashi. "Geothermal Energy for Electric Power." Renewable Energy Policy Project. Available online. URL: http://solstice.crest.org/articles/static/1/binaries/Geothermal_Issue_Brief.pdf. Posted in December 2003. Provides background on utility-scale geothermal energy and the technical, economic, and policy dimensions of its worldwide development.

"Solar Electricity in 2010." European Photovoltaic Industry Association. Available online. URL: http://www.epia.org/documents/Solar_Electricity_2010.pdf. Downloaded on July 22, 2004. Includes solar energy as a key to a sustainable energy future, technology options and advantages, market

projections for the solar electricity industry, and strategies for fostering investment in solar electricity in Europe.
"Solar Energy Types." Solar Energy Industries Association. Available online. URL: http://www.seia.org/learn/energytypes.asp. Downloaded on July 23, 2004. Describes photovoltaics, concentrating solar power (large solar power plants that do not use photovoltaics), and solar water heating.
"Solar Generation." European Photovoltaic Industry Association. Available online. URL: http://www.epia.org/03publications/publications.htm#. Leaflets. Downloaded on July 22, 2004. Covers solar basics, the power market, job creation, winners and losers in solar generation, and the future of solar power.
"Solar Myths and Facts." Solar Energy Industries Association. Available online. URL: http://www.seia.org/learn/myths.asp. Downloaded on July 23, 2004. Provides evidence to counter alleged "myths" such as the idea that solar power is too expensive for widespread use.
"Sun in Action II: A Solar Thermal Strategy for Europe." European Solar Thermal Industry Foundation. Available online. URL: http://www.estif.org/fileadmin/downloads/sia/SiA2_Vol1_final.pdf. Posted in May 2003. Two volumes, one that surveys the European market in general and one that reports on markets in specific countries. The web page provides an action plan to tackle obstacles to growth and stresses the need for continued government support.
"The Ten-Point Plan for Good Jobs and Energy Independence." Apollo Alliance. Available online. URL: http://www.apolloalliance.org/strategy_center/a_bold_energy_and_jobs_policy/ten_point_plan.cfm. Downloaded on July 22, 2004. Stresses the economic value of increased energy efficiency and use of renewable energy sources.
"United States RPS Case Studies." Center for Resource Solutions. Available online. URL: http://www.resource-solutions.org/Library/librarypdfs/IntPolicy-USRPSCASESTUDIES.pdf. Downloaded on July 22, 2004. Compares and analyzes Renewable Portfolio Standard programs in Texas, Wisconsin, and Maine and summarizes lessons learned from them.
"Why Hydropower Should Be Included in Renewable and Sustainable Energy Initiatives." National Hydropower Association. Available online. URL: http://www.hydro.org/pdf/FRIHI.pdf. Downloaded on July 23, 2004. Offers 12 reasons why hydropower should be classified as a renewable/sustainable rather than conventional energy source.
"Why Solar?" Vote Solar Initiative. Available online. URL: http://www.votesolar.org/whysolar.html. Downloaded on July 23, 2004. Gives reasons why city governments should implement solar projects.
"Wind Energy—The Facts." European Wind Energy Association. Available online. URL: http://www.ewea.org/06projects_events/proj_WEfacts.htm.

Posted in 2003. Comprehensive overview of the wind energy sector's past, present, and future in the European Union. Five volumes, covering technology, costs and prices, industry and employment, environment, and market development.

"Wind Force 12." European Wind Energy Association. Available online. URL: http://www.ewea.org/documents/WF12–2004_eng.pdf. Posted in May 2004. A blueprint to achieve 12 percent of the world's energy from wind power by 2020.

UTILITIES AND THEIR REGULATION

BOOKS

Beder, Sharon. *Power Play: The Fight to Control the World's Electricity.* New York: New Press, 2003. Shows how corporations seized control of electricity from public hands and are now pushing for deregulation. Events in Brazil and California make the author pessimistic about deregulation's outcome.

Borbely, Anne-Marie, and Jan F. Kreider, eds. *Distributed Generation: The Power Paradigm for the New Millennium.* Boca Raton, Fla.: CRC Press, 2001. Surveys technologies for small-scale electricity generation and argues that creating energy at or near its point of use is better than the present system using huge, centralized power plants and long-distance transmission.

Brennan, Timothy J., et al. *A Shock to the System: Restructuring America's Electricity Industry.* Washington, D.C.: Resources for the Future, 1996. Includes the industry's regulatory history, different models of competition, pros and cons of vertical integration, and issues of stranded costs and environmental protection.

Brennan, Timothy J., Karen L. Palmer, and Salvador A. Martinez. *Alternating Currents: Electricity Markets and Public Policy.* Washington, D.C.: Resources for the Future, 2002. Describes the issues involved in deciding whether, when, and how to open electricity markets to competition. The book includes discussion of market power, features that make electricity a unique resource, and the California crisis.

Carlson, John, and G. A. Oyibo, eds. *Electricity Restructuring: Issues and Policy Questions.* Hauppauge, N.Y.: Nova Science Publishers, 2002. Weighs deregulation against concerns relating to air pollution and consumer rights and concludes that a hybrid restructuring is more desirable than total deregulation.

Chambers, Ann. *Natural Gas and Electric Power in Nontechnical Language.* Tulsa, Okla.: Pennwell Publishing, 1999. Describes the history of the electricity and natural gas industries, focusing on the political, regulatory, economic, and technical factors that caused them to converge.

Cicchetti, Charles J., Jeffrey A. Dubin, and Colin M. Long, eds. *The California Electricity Crisis: What, Why, and What's Next.* New York: Kluwer Academic Publishers, 2004. Analysts comment on the factors that led to the 2000–01 crisis, the events of the crisis, and their implications for electricity deregulation.

Energy Information Administration. *Public Utility Holding Company Act of 1935: 1935–1992.* Booklet describes this law, arguments for and against reforming it, and its modification by the Public Utility Regulatory Policies Act (1978) and the Energy Policy Act (1992).

Ferrey, Steven. *The New Rules: A Guide to Electric Market Regulation.* Tulsa, Okla.: Pennwell Publishig, 2001. Describes the new, liberalized deregulatory environment in the areas of technology, federal regulation, and retail sales.

Geradin, Damien, ed. *The Liberalization of Electricity and Natural Gas in the European Union.* New York: Kluwer Law International, 2001. Energy experts consider issues raised by liberalization of the electricity and natural gas markets in the European Union in competition, transmission and trading, and environment and consumer protection. They describe the national experiences of Belgium, France, Germany, and the Netherlands.

Glachant, Jean-Michel, and Dominique Finon, eds. *Competition in European Electricity Markets: A Cross-Country Comparison.* Northampton, Mass.: Edward Elgar, 2003. Shows the diversity of electricity reforms in Western Europe by describing the actions of 12 countries, with a focus on economics. Offers explanations for the differences in the countries' approaches to deregulation.

Hampton, Howard. *Public Power: The Fight for Publicly Owned Electricity.* Toronto, Ontario, Canada: Insomniac Press, 2003. Describes how the North American electricity industry has evolved during the last century, focusing on the struggle between private and public power, the political origins of the current deregulation movement, and case studies showing why the author thinks deregulation is a failure. Hampton argues for a regulated, publicly owned power system.

Hirsch, Richard F. *Power Loss: The Origins of Deregulation and Restructuring in the American Electric Utility System.* Cambridge, Mass.: MIT Press, 2000. Describes how and why the U.S. electric utility industry was radically restructured in the late 1990s to increase competition, including interactions between technological change and regulation.

Horsnell, Paul, Gordon Mackerron, and Peter Pearson, eds. *UK Energy Policy and the World Market.* New York: Oxford University Press, 2003. Papers from a 1999 conference discuss regulation of energy markets in Britain.

Hunt, Sally. *Making Competition Work in Electricity.* Hoboken, N.J.: John Wiley & Sons, 2002. Explains what needs to be accomplished for competition in electricity markets to work in any country, then describes the current complicated situation in the United States and shows how author's "standard prescription" can be applied to it.

Hyman, Leonard S., Andrew S. Hyman, and Robert C. Hyman. *America's Electric Utilities: Past, Present, and Future.* 7th ed. Vienna, Va.: Public Utilities Reports, 2000. Clearly explains technical, business, and regulatory aspects.

Jonnes, Jill. *Empires of Light: Edison, Tesla, Westinghouse and the Race to Electrify the World.* New York: Random House, 2003. Describes the beginnings of the U.S. electricity industry, focusing on its three chief competing inventors and the financial forces that controlled them all.

Krishnaswamy, Venkataraman, and Gary Stuggins. *Private Participation in the Power Sector in Europe and Central Asia: Lessons from the Last Decade.* Washington, D.C.: World Bank Group, 2003. Draws four key lessons from the 1990s electricity sector experiences of Hungary, Lithuania, Poland, and Turkey as well as six former states of the Soviet Union, and Russia, Romania, Armenia, and Albania.

Lai, Loi Lei, ed. *Power System Restructuring and Deregulation.* Hoboken, N.J.: John Wiley & Sons, 2001. Examines the deregulated electricity landscape worldwide through case studies and modeling, including competitive, regulatory, and technological changes, and considers how these changes affect the management of energy utility companies.

Lambert, Jeremiah D. *Creating Competitive Power Markets: The PJM Model.* Tulsa, Okla.: Pennwell Publishing, 2001. Describes PJM, the Independent System Operator (ISO) organization that manages electrical transmission for Pennsylvania, New Jersey, and Maryland, and compares it favorably with other ISOs.

Lovins, Amory. *Small Is Profitable: The Hidden Economic Benefits of Making Electrical Resources the Right Size.* Snowmass, Colo.: Rocky Mountain Institute, 2002. Tells why today's electrical industry, featuring large central power stations and a nationwide transmission grid, soon will—and, for good economic reasons, should—be replaced by a decentralized, distributed network.

MacAvoy, Paul W. *The Natural Gas Market: Sixty Years of Regulation and Deregulation.* New Haven, Conn.: Yale University Press, 2001. Describes federal regulatory strategies that have been applied to the industry in the United States during the past 60 years. MacAvoy concludes that all have been unsuccessful and argues that complete deregulation would be better.

Mackerron, G., and Luigi de Paoli, eds. *The Electricity Supply Industry of Europe.* London: Earthscan Publications, 2003. Describes legal, regulatory, economic, commercial, and social aspects of the industry in each of the 12 European Union member countries, focusing on the effects of the deregulation required by the EU Electricity Directive of 1997.

McNamara, Will. *The California Energy Crisis: Lessons for a Deregulating Industry.* Tulsa, Okla.: Pennwell Publishing, 2002. Presents the facts of California's deregulation scheme and the subsequent 2000–01 crisis, discusses blame, describes the crisis's impact on California's and the nation's economy and policy, and draws lessons from it that other states should heed.

Middtun, Atle, ed. *European Energy Industry Business Strategies.* St. Louis, Mo.: Elsevier Health Sciences, 2001. Describes the European Union's newly deregulated energy industry and business strategies that have emerged within it.

Newbery, David M. *Privatization, Restructuring, and Regulation of Network Utilities.* Cambridge, Mass.: MIT Press, 2000. Examines how relationships among consumers, investors, and regulators affect network utilities such as electricity, natural gas, and telecommunications suppliers. Newbery compares regulatory approaches in Britain and the United States; he also briefly describes those of other countries.

Ocana, Carlos, ed. *Competition in Electricity Markets.* Paris: Organisation for Economic Cooperation and Development, 2001. Draws on experiences in many countries to consider the nature of an effective regulatory framework for competition in electricity markets.

O'Donnell, Arthur J. *Soul of the Grid: A Cultural Biography of the California Independent Systems Operator.* Lincoln, Neb.: iUniverse.com, 2003. This independent transmission organization was both praised and criticized for its role in the 2000–01 California energy crisis. The book tells how it came to be and what happened during its first five years; it also describes the people who make up the organization's culture and "soul."

Qayoumi, Mohammad H., ed. *Changing Currents in Deregulation.* Washington, D.C.: APPA, 2001. Describes past, present, and predicted future regulatory conditions in the U.S. electricity industry, including effects on pricing and availability.

Savage, Norton. *An Electric Power System Vocabulary.* Raleigh, N.C.: Ivy House Publishing, 2003. More than 60 entries exploring every aspect of electrical power systems, including excerpts from technical papers.

Smith, Rebecca, and John R. Emshwiller. *24 Days: How Two* Wall Street Journal *Reporters Uncovered the Lies that Destroyed Faith in Corporate America.* New York: HarperCollins, 2003. A reporter covering the energy industry just after the California crisis teamed with a second one covering stock swindlers to expose the secrets of Enron's market manipulation.

Vrolijk, Christiaan, ed. *Climate Change and Power: Economic Instruments for European Electricity.* London: Earthscan Publications, 2003. Describes the characteristics of the major European electricity sectors and the economic policy instruments in use or likely to be developed to control carbon emissions from electricity generation.

Warkentin, Denise. *Electricity Power Industry in Nontechnical Language.* Tulsa, Okla.: Pennwell Publishing, 1998. Concise explanation of the history of the industry, how it works, energy sources, laws and regulations, and economic issues affecting it.

Wasserman, Harvey. *The Last Energy War: The Battle over Utility Deregulation.* New York: Seven Stories Press, 2000. Argues that utility deregulation threatens consumers.

Weare, Christopher. *The California Electricity Crisis: Causes and Policy Options.* San Francisco: Public Policy Institute of California, 2003. Shows how multiple factors combined to cause the crisis, discusses options for rebuilding the state's electricity sector, and offers recommendations for improving sector performance under any regulatory and market structure.

Wellstone, Paul, and Barry M. Casper. *Powerline: The First Battle of America's Energy War.* Minneapolis: University of Minnesota Press, 2003. Describes a battle that Minnesota farmers fought in the late 1970s against the building of a high-voltage power line from central North Dakota to the Twin Cities suburbs, based on the belief that the line threatened the health of their families and land.

Xu, Yi-Chong. *Powering China: Reforming the Electric Power Industry in China.* Hampshire, U.K.: Dartmouth Publishing Co., 2002. Examines the complex relationships between politics and economics that underlie the restructuring of China's electricity industry, showing that electricity reform is part of a general shift from central planning to a market economy.

ARTICLES

Alexander, Santosh. "Lessons from California." *Energy*, vol. 26, Fall 2001, pp. 17–20. How California's 2000–01 energy crisis occurred and what other states considering electricity deregulation can learn from it.

Arjan, Asthana, Kenneth J. Ostrowski, and Bangalore Venkateshwara. "Finding the Balance of Power." *The McKinsey Quarterly*, Autumn 2001, pp. 82 ff. Utility deregulation has spawned new breeds of generators and power traders.

Betz, Kenneth W. "Enron's Financial Fallout Is Clear, but Deregulation's Future Is Partly Cloudy." *Energy User News*, vol. 27, February 2002, pp. 1–2. The collapse of Enron does not necessarily mean that deregulation/restructuring the electricity industry is a bad idea.

Blankinship, Steve. "Distributed Generation: DGenie Is Out of the Bottle." *Power Engineering*, vol. 107, March 2003, pp. 26–30. Technological advances and economic incentives are encouraging industrial and other electricity customers to consider generating their own electricity, especially when doing so can be coupled with other industrial processes.

Burkhart, Lori A. "FERC Addresses California Crisis, Enrages Gov. Davis." *Public Utilities Fortnightly*, vol. 141, August 2003, pp. 12–13. The Federal Energy Regulatory Commission apportions blame and punishment to Enron and others for their behavior in the 2000–01 California energy crisis but does not let the state escape the long-term power contracts it signed at the time.

Carvazos, Roberto J., and Terry F. Buss. "Electric Industry Restructuring: An Overview of the Policy Issues." *The Review of Policy Research*, vol. 20, pp. 203–217. Lists policy issues involved in electricity generation and provides detailed analysis of the 2000–01 California energy crisis.

"Charged with Controversy." *Canada and the World Backgrounder*, vol. 66, March 2001, pp. 8–9. Electricity deregulation in Canada has gotten off to a shaky start, thanks to government mismanagement.

"The Dawn of Micropower." *The Economist*, vol. 356, August 5, 2000, p. 75. Micropower, or distributed generation, is growing cheaper, thanks to technological innovation, and is likely to become increasingly popular.

Easton, Bill. "Deregulation Diary." *Utility Week*, July 11, 2003, p. 26. Describes changes in European electricity markets in the second half of 2002 in response to increased deregulation.

Ferrey, Steven. "Inverting Choice of Law in the Wired Universe: Thermodynamics, Mass, and Energy." *William and Mary Law Review*, vol. 45, April 2004, pp. 1839–1956. Discusses the restructuring of the California electricity market, including relevant laws. Focuses on the question of whether electricity is a good or a service and concludes that, although it has some features of each, physical reality makes it mostly a service.

Foroohar, Rana, Ian MacKinnon, and Mac Margolis. "What Not to Do." *Newsweek International*, April 8, 2002, p. 51. Lessons the European Union has learned from energy mistakes in the United States.

Gabaldon, Dan, and Joe Quoyeser. "Vertical Integration in Gas and Power: Necessity or Distraction?" *Public Utilities Fortnightly*, vol. 140, April 15, 2002, pp. 28–33. Describes two alternative future scenarios for the U.S. natural gas and electricity industries and lists indicators that may show which scenario is more likely to develop.

Gorman, Robert O. "Significant Trends in the U.S. Electric Power Industry." *Power, Finance and Risk*, vol. 5, November 4, 2002, pp. 6–7. The electricity industry in the United States is in turmoil, but several important trends can be discerned.

Hazan, Earl, and Pam Kufahl. "The Quest for Power." *Transmission & Distribution World*, December 2000, n.p. Advanced by a host of creative innovators, the electricity industry transformed society in the 20th century.

Hirsh, Michael, and Daniel Klaidman. "What Went Wrong." *Newsweek*, August 25, 2003, pp. 32–33. Vividly describes the Northeast blackout of August 14, 2003, the worst in U.S. history, and considers its causes.

Howe, John B. "Edison Shrugged." *Public Utilities Fortnightly*, vol. 141, September 1, 2003, pp. 15–17. The hybrid of regulation and market forces that presently controls the U.S. electricity industry creates uncertainties that discourage investment in transmission infrastructure, threatening the system's reliability and the health of the American economy.

Isser, Steven N. "Electricity Deregulation: Kilowatts for Nothing and Your BTUs for Free." *The Review of Policy Research*, vol. 20, pp. 219–238. Provides history of deregulation of natural gas and electricity in the United States beginning in the 1970s and discusses why deregulation of retail electricity markets has not brought promised benefits to consumers.

Karmali, Abyd, and Myfanwy Price-Jones. "Carbon and the Strategic Implications of Emission Limits on European Power." *Power Economics*, vol. 6, July–August 2002, pp. 16–17. Electricity generation patterns in Europe are changing rapidly, and European Union plans for reducing greenhouse gas emissions will exacerbate the economic effects of those changes.

Kavanagh, Ronan. "Wired World Places Strain on Power Grids." *Energy Network*, vol. 3, April 9, 2002, p. 3. The worldwide rise in use of computer technology is adding significantly to energy consumption—and the power grids of many countries are not ready to supply what is needed.

"Keeping the Juice Flowing." *Business Week*, March 4, 2002, p. 94B. New software can help to fix problems with the reliability of the electric transmission grid.

Kluger, Jeffrey. "Getting By without the Grid." *Time*, vol. 162, August 25, 2003, p. 36. Distributed generation, in which clusters of homes produce most of their own electricity, will probably supplement the centralized electricity grid in the future, but experts disagree about how much of a role they will play.

Ladd, Conrad M. "Power to the People." *Mechanical Engineering-CIME*, vol. 122, September 2000, pp. 68 ff. Inventive engineers have helped the electric power industry develop and meet challenges during the 20th century.

Lesser, Jonathan A., and Charles D. Feinstein. "Distributed Generation: Hype vs. Hope." *Public Utilities Fortnightly*, vol. 140, June 1, 2002, pp. 20–25. The benefits of distributed generation are real, but they have sometimes been exaggerated, leading to application in inappropriate situations. Economic analysis shows that the benefits are greatest when uncertainty about load growth is large and the expected rate of growth is small.

Liddy, Glen. "A New Electric Economy: Powering the Future?" *Power Economics*, vol. 5, June 2001, pp. 18 ff. Discusses predicted changes in the means of generating and distributing electricity in the next 50 years, including the roles of renewable energy sources and distributed generation.

MacFarlane, Derek. "Toward the Integrated Utility Network." *Power Economics*, vol. 6, September 2002, pp. 26–27. An integrated information system is essential for preventing electricity reliability problems and increasing efficiency. MacFarlane includes analysis of an extensive power failure in Auckland, New Zealand, in 1998.

Marshall, Christine, et al. "What FERC Orders Really Mean." *Energy User News*, vol. 27, January 2002, pp. 10–13. The Federal Energy Regulatory Commission's attempts to have transmission of electricity in the United States overseen by only four large regional organizations is likely to produce instability in the short term but will benefit consumers in the long run.

Mattoon, Richard. "The Electricity System at the Crossroads—Policy Choices and Pitfalls." *Economic Perspectives*, vol. 26, Spring 2002, pp. 2–18. States' experiences of electricity deregulation have been varied. Important lessons can be learned from those experiences about what factors and objectives should be most important in determining policy.

McCarthy, Elizabeth. "Keeping Reporters and the Public in the Dark." *Nieman Reports*, vol. 58, Summer 2004, pp. 22–24. "Rolling information blackouts" at both state and federal levels and secret dealmaking hampered reporters trying to tell the public what was happening during the 2000–01 California energy crisis.

Ockenden, Karma. "The Pace of Change." *Utility Europe*, August 1, 2002, p. 22. Energy companies will need to change in several ways to stay competitive in the future.

"Power to the Poor." *The Economist*, February 10, 2001, p. 8. Small local projects and market-based incentives can help to bring electricity to the world's poorest people.

Russell, Naomi. "The Energy Within." *Utility Europe*, August 1, 2002, p. 16. Summarizes arguments for and against a predicted increase in distributed electricity generation in Europe.

Speakes, Kelly. "Dispersed Generation." *Power Economics*, vol. 8, February 2004, p. 12. "Dispersed generation" will not completely remove bottlenecks from the U.S. power grid, but it could help.

Stone, Brad. "How to Fix the Grid." *Newsweek*, August 25, 2003, p. 38. Suggests ways to improve the U.S. electricity transmission system.

Tackett, Matthew H. "How to Minimize Cascading Event Probabilities." *Electric Light & Power*, vol. 81, September 2003, p. 6. Recommends regulatory, operational, and planning reforms to improve the reliability of the electric transmission grid.

Timney, Mary M. "Short Circuit: Federal-State Relations in the California Energy Crisis." *Publius*, vol. 32, Fall 2002, pp. 109–123. Examines the changing roles of state and federal governments in determining electricity regulatory policy, with a focus on California. Timney maintains that these changes have increased federal responsibility, which the Federal Energy Regulatory Commission failed to fulfill during the state's energy crisis.

Tomain, Joseph P. "The Past and Future of Electricity Regulation." *Environmental Law*, vol. 32, Spring 2002, pp. 435–474. Argues that restructuring the electricity sector is sound industrial policy, but the industry can never be completely deregulated because electricity transmission, unlike generation, is a natural monopoly.

Whitlow, David. "How to Avoid the Nightmare Scenario." *European Power News*, vol. 29, April 2004, pp. 27–28. Explains how power outages happen and how to reduce their risk, with a focus on Western Europe.

WEB DOCUMENTS

Abel, Amy. "Electricity Restructuring Background: The Public Utility Regulatory Policies Act of 1978 and the Energy Policy Act of 1992." National Council for Science and the Environment. Available online. URL: http://www.ncseonline.org/nle/crsreports/energy/eng-36.cfm. Posted on May 4, 1998. Congressional Research Service report describes these two laws and their effects, as well as the modifications that the Energy Policy Act created in an earlier law, the Public Utility Holding Company Act.

Besant-Jones, John, and Bernard Tenenbaum. "The California Power Crisis: Lessons for Developing Countries." World Bank. Available online. URL: http://rru.worldbank.org/Documents/PapersLinks/365.pdf. Posted in April 2001. Warns that developing countries should not try to introduce widespread competition in electricity markets too quickly. Instead, they should concentrate on charging and collecting tariffs that cover costs.

"Competition Works in Electric Power Markets: 2004 Update." Electric Power Supply Association. Available online. URL: http://www.epsa.org/forms/uploadFiles/37F400000002.filename.CompWorksBR_5.pdf. Posted on March 15, 2004. Brochure with tables and charts showing falling prices for electricity, allegedly because of competition.

"Essential Elements of Well Functioning Competitive Power Markets." Electric Power Supply Association. Available online. URL: http://www.epsa.org/forms/uploadFiles/35B300000020.filename.eewellfunction.pdf.

Posted in March 2004. Describes features including market scope, governance and monitoring, and transmission management.

Farmer, Richard, Dennis Zimmerman, and Gail Cohen. "Causes and Lessons of the California Electricity Crisis." Congressional Budget Office. Available online. URL: http://www.cbo.gov/showdoc.cfm?index=3062&sequence=0. Posted in September 2001. Describes conditions in the western states that stressed California's energy market in 2000 and features of the state's deregulation plan that exacerbated that stress into a crisis, including rate caps and a provision forbidding long-term contracts.

Harvey, Hal, Bentham Paulos, and Eric Heitz. "California and the Energy Crisis: Diagnosis and Cure." Energy Foundation. Available online. URL: http://www.ef.org/california/downloads/CA_crisis.pdf. Posted on March 8, 2001. Blames the crisis on changes in the fundamentals of the western electric power situation and mistakes in designing new markets. The authors stress energy efficiency and renewable energy sources as solutions.

"National Transmission Grid Study." U.S. Department of Energy. Available online. URL: http://tis.eh.doe.gov/ntgs/reports.html. Posted in 2001. Stresses the need to repair the nation's aging electricity transmission infrastructure and upgrade it to include improved technology that could increase electricity flow through the system, such as high-temperature superconducting cables.

Sell, T. M. "What If You Deregulated a Market, and No One Cared?" Salon.com. Available online. URL: http://archive.salon.com/tech/feature/2004/05/24/electricity_deregulation. Posted on May 24, 2004. Claims that even in states where electricity deregulation supposedly has been successful, such as Pennsylvania, consumers still have little choice about their electricity suppliers and do not save much money.

Swisher, Joel N. "Cleaner Energy, Greener Profits: Fuel Cells as Cost-Effective Distributed Energy Resources." Rocky Mountain Institute. Available online. URL: http://www.rmi.org/images/other/Energy/U02-02_ CleanerGreener.pdf. Posted in 2002. Claims that a predicted turn toward distributed generation will make fuel cells cost effective.

U.S.-Canada Power System Outage Task Force. "Final Report on the August 14, 2003 Blackout in the United States and Canada: Causes and Recommendations." Natural Resources Canada. Available online. URL: http://www.nrcan-rncan.gc.ca/media/docs/final/BlackoutFinal.pdf. Posted in April 2004. Includes overview of the North American electric power system and its reliability organizations, causes of the 2003 northeastern blackout, condition of the power grid before the blackout, how and why the blackout began, stages of the blackout, and recommendations for preventing or minimizing future blackouts.

GLOBAL ENERGY ISSUES

BOOKS

Aditjondro, George J. *Is Oil Thicker than Blood?: A Study of the Oil Companies' Interests and Western Complicity in Indonesia's Nexation of East Timor.* Hauppauge, N.Y.: Nova Science Publishers, 1999. Claims that eagerness to obtain control of East Timor's offshore and onshore oil and gas reserves was the main reason why Indonesia took over this area, with the tacit approval of Western nations and U.S. oil interests.

Alnasrawi, Abbas. *Iraq's Burdens: Oil, Sanctions, and Underdevelopment.* Westport, Conn.: Greenwood Publishing Group, 2002. Shows how oil revenue has been a curse for Iraq and, by implication, is so for other oil-rich but otherwise underdeveloped countries as well.

Aminehj, Mehdi Parvizi. *Toward the Control of Oil Resources in the Caspian Region.* New York: Palgrave Macmillan, 2000. Focuses on the post-Soviet geopolitics of this important area and the cooperation and conflict among regional powers, multinational corporations, and Western powers as these groups seek to control the region's oil and gas resources.

Amirahmadi, Hooshang, ed. *The Caspian Region at a Crossroads: Challenges of a New Frontier of Energy and Development.* New York: Palgrave Macmillan, 2000. Essays by experts correct some misunderstandings about this newly important oil-and-gas-rich region.

Amuzegar, Jahangir. *Managing the Oil Wealth: OPEC's Windfalls and Pitfalls.* New York: I. B. Tauris, 2001. Analyzes factors that led to OPEC's rise and alleged decline and discusses why countries so different in many respects fell into the same "oil trap" and wasted the money they obtained from this resource.

Bacher, John. *Petrotyranny.* Toronto, Ontario, Canada: Science for Peace/Samuel Stevens, 2000. Points out that most of the world's oil resources are controlled by dictators, creating a great potential for global conflict as oil supplies dwindle. Bacher maintains that the world's democracies can avoid these problems by switching to renewable energy sources.

Baer, Robert. *Sleeping with the Devil: How Washington Sold Our Soul for Saudi Crude.* New York: Crown, 2003. Claims that the close relationship between the United States and Saudi Arabia is hypocritical and corrupt and indirectly fosters terrorism and potential war.

Belyaev, Lev S., et al., eds. *World Energy and Transition to Sustainable Development.* New York: Kluwer Academic Publishers, 2002. Members of the Russian Academy of Sciences describe studies on the effects of trends in technological progress on the development of national energy sectors around the world. The book discusses requirements for sustainable development and explores effects of mixing energy technologies.

Bloomfield, Lincoln P., Jr., and James A. Kelly, Jr., eds. *Global Markets and National Interests: The New Geopolitics of Energy, Capital, and Information.* Washington, D.C.: Center for Strategic and International Studies, 2002. Describes energy geopolitics and shows ways in which globalization is affecting U.S. foreign policy.

Brisard, Jean-Charles, and Guillaume Dasquié. *Forbidden Truth: U.S.-Taliban Secret Oil Diplomacy and the Failed Hunt for Bin Laden.* New York: Thunder's Mouth Press, 2002. Claims that the Clinton and Bush administrations tried to stabilize Afghanistan so U.S.-based energy companies could build a pipeline through the country.

Brown, Anthony Cave. *Oil, God, and Gold: The Story of Aramco and the Saudi Kings.* Boston: Houghton Mifflin, 1999. Describes complex relationships among the Saudi royal family, multinational oil executives, and the American and British governments that emerged in the 1930s and became intense after World War II. The book focuses on Aramco (originally the Arabian American Oil Company) and draws heavily on documents from that company.

Brown, Charles E. *World Energy Resources.* New York: Springer Verlag/Telos Press, 2002. Provides an in-depth, country-by-country analysis of global energy resources and international energy markets in the past, present, and future. The book includes discussion of transportation, electricity, and environmental issues.

Campbell, Charles L. *On a Clear Day I Can See Armageddon: The Collision of Israel, Islam, and Oil.* Tucson, Ariz.: Fenestra Books, 2003. Describes the religious and political history of the Middle East and the current situation there, especially in Iraq. Campbell offers policy, recommendations.

Chalabi, Fadhil, and Muhammad-Ali Zainy. *Post-Saddam Iraq: Oil and Gas, Economy, Finance, Politics.* London: Centre for Global Energy Studies, 2004. Provides historical, financial, and political background on Iraq and discusses the past, present, and future prospects of the country's oil industry.

Citino, Nathan J. *From Arab Nationalism to OPEC: Eisenhower, King Sa'ud, and the Making of U.S.-Saudi Relations.* Bloomington: Indiana University Press, 2002. Reexamines the relationship between President Dwight Eisenhower and the founder of Saudi Arabia after World War II, the United States's replacement of Britain as the leading foreign power in the Middle East, and the formation of OPEC, which the author claims was a fulfillment of U.S. goals in the Middle East.

Claes, Dag Harald. *The Politics of Oil-Producer Cooperation.* Boulder, Colo.: Westview Press, 2001. Extensive study of political actors in the world oil market since 1971 attempts to find out what determines cooperative behavior among oil-producing countries.

Cummings, Sally N., ed. *Oil, Transition and Security in Central Asia.* Richmond, Surrey, U.K.: Curzon Press, 2003. Provides a broad, multidisciplinary analysis of domestic and international developments in the region since 1991.

Day, James McDonald. *What Every American Should Know About the Mid East and Oil.* Carson City, Nev.: Bridger House Publishing, 2001. Strongly opposes the George W. Bush administration's Middle East policies, which the author says are based on oil interests.

Doig, Alison, et al. *Energy for Rural Livelihoods: A Framework for Sustainable Decision Making.* Rugby, Warwickshire, U.K.: ITDG Publishing, 2004. Provides guidance for integrating the complex factors involved in bringing sustainable energy services to rural communities into a coherent framework for decision making.

Downs, Erica Strecker. *China's Quest for Energy Security.* Santa Monica, Calif.: Rand Corporation, 2000. Describes measures China is taking to try to achieve energy security while facing the fact that it must increasingly depend on imported sources as its demand rises.

Doyle, Jack. *Riding the Dragon: Royal Dutch/Shell and the Fossil Fire.* Monroe, Maine: Common Courage Press, 2004. Claims that the company's projects around the world have harmed health and the environment.

Ebel, Robert, and Rajan Menon, eds. *Energy and Conflict in Central Asia and the Caucasus.* Lanham, Md.: Rowman & Littlefield, 2000. Examines the relationship between tendencies for conflict in the Caspian area and competition for energy resources, especially oil and natural gas. The book places the Caspian in the larger context of international politics.

Everest, Larry. *Oil, Power and Empire: Iraq and the U.S. Global Agenda.* Monroe, Maine: Common Courage Press, 2003. Claims to reveal the hidden, oil-related agenda behind the U.S. invasion of Iraq in 2003.

Farmanfarmaian, Manucher, and Roxane Farmanfarmaian. *Blood and Oil: Memoirs of a Persian Prince.* New York: Random House, 1997. Former Iranian prince and director of the National Iranian Oil Company describes the end of British domination of the Middle East and emergence of the oil industry and control by the United States.

Gökay, Bülent, ed. *The Politics of Caspian Oil.* New York: Palgrave Macmillan, 2001. Essays on many aspects of this complex subject.

Haley, James, ed. *At Issue: Foreign Oil Dependence.* San Diego, Calif.: Greenhaven Press, 2004. Anthology of essays provides a range of viewpoints on issues related to U.S. dependence on oil imports.

Heinberg, Richard. *The Party's Over: Oil, War and the Fate of Industrial Societies.* Gabriola Island, B.C., Canada: New Society Publications, 2003.

Places in a historical context the world's upcoming forced transition away from fossil fuels as these fuels become scarce and expensive. Author predicts chaos unless the United States joins other countries to implement a program of resource conservation and sharing in order to prevent "resource wars" over remaining fossil fuel supplies.

Horsnell, Paul. *The Mediterranean Basin in the World Petroleum Market.* New York: Oxford University Press, 1999. Considers development of the oil and oil refining industries in this area, including Iraq and the Caspian Sea region; stresses the area faces; and its place in the world oil market.

Kayal, Alawi D. *The Control of Oil.* London: Kegan Paul, 2002. The author, a former Saudi Arabian minister, examines the 20th-century struggle among the United States, the U.S.S.R., and other countries for control of Arab oil fields, the issues and events leading to the formation of the Organization of Petroleum Exporting Countries (OPEC), and changes that followed through the 1970s.

Kerr, William A., and Jennifer I. Considine. *The Russian Oil Economy.* Northampton, Mass.: Edward Elgar, 2002. Reviews the history of Russian oil industry planning and the interaction between Russian and international oil markets. The book stresses the likely future importance of Russia in the world oil market.

Klare, Michael T. *Resource Wars: The New Landscape of Global Conflict.* New York: Henry Holt, 2001. Predicts that wars will increasingly be fought over scarce resources, especially fossil fuels and water, as growing populations make heavier demands on them.

Kleveman, Lutz. *The New Great Game: Blood and Oil in Central Asia.* Boston: Atlantic Monthly Press, 2003. Focuses on competition among the United States, Russia, China, and multinational oil companies for control of the oil wealth of the Caspian Sea region.

Lane, David Stuart, ed. *The Political Economy of Russian Oil.* Lanham, Md.: Rowman & Littlefield, 1999. Uses Russia's oil industry as a case study to explore political and economic issues surrounding the country's move toward capitalism.

Manibog, Fernando, Rafael Dominguez, and Stephen Wegner. *Power for Development: A Review of the World Bank Group's Experience with Private Participation in the Electricity Sector.* Washington, D.C.: World Bank, 2004. Evaluates the performance of this international lending institution during the 1990s in 80 countries. The authors conclude that the bank should continue to support private participation in electricity, but the outcome of many of its projects was disappointing.

Manning, Robert A. *The Asian Energy Factor: Myths and Dilemmas of Energy, Security and the Pacific Future.* New York: Palgrave Macmillan, 2000. Assesses energy challenges and strategies of Asian nations. The author maintains that political and economic bottlenecks, not actual shortages of fuel supplies, cause most current conflicts and problems.

Mommer, Bernard, and Ali Rodriguez Araque. *Global Oil and the Nation State.* New York: Oxford University Press, 2002. Analyzes long-term relationships among natural resource owners, investors, and consumers, focusing on the political and institutional dimensions of the oil economy.

National Research Council, Chinese Academy of Sciences, and Chinese Academy of Engineering. *Cooperation in the Energy Futures of China and the United States.* Washington, D.C.: National Academies Press, 2000. Describes linkage between the two countries in the area of energy and discusses environmental consequences of energy production and use.

Noreng, Oystein. *Crude Power: Politics and the Oil Market.* New York: I. B. Tauris, 2002. Intensive analysis of the world's dependency on Middle Eastern oil.

Nunn, Sam, and James R. Schlesinger, eds. *The Geopolitics of Energy into the 21st Century. Volume I: An Overview and Policy Considerations.* Washington, D.C.: Center for Strategic and International Studies, 2002. Claims that the relative calm in world energy markets since 1985 masks highly significant shifts in energy geopolitics that U.S. foreign policy is ignoring.

Odell, Peter. *Oil and Gas: Crises and Controversies, 1961–2000.* Brentwood, Essex, U.K.: Multi-Science Publishing Co., Ltd., 2004. Two volumes. Volume I deals with global issues, while Volume II focuses on Europe. Odell describes how the world oil market and international politics have changed one another.

Olcott, Martha Brill, ed. *Kazakhstan: Unfulfilled Promise.* Washington, D.C.: Carnegie Endowment for International Peace, 2002. Describes corruption and internal dissension in this oil-rich former Soviet nation during its first decade of independence.

Omoweh, Daniel A. *Shell Petroleum Development Company, the State and Underdevelopment of Nigeria's Niger Delta.* Trenton, N.J.: Africa World Press, 2004. Claims that the oil company's activities in this African nation produced severe environmental and social damage.

Parra, Francisco. *Oil Politics: A Modern History of Petroleum.* New York: I. B. Tauris, 2004. Describes the political and economic events that have shaped the international petroleum industry over the past half century, stressing the triangular relationship between Middle Eastern governments, the major international oil companies, and the governments of the companies' home countries (the United States and Britain).

Peimani, Hooman. *The Caspian Pipeline Dilemma: Political Games and Economic Losses.* New York: Praeger, 2001. Claims that the American policy of isolating Iran cuts off the best pipeline routes from the oil-and-gas-rich, landlocked Caspian Sea area, limiting development of the area and threatening to cut off the United States from the bounty of this region if, as the author predicts, oil and gas exporters build a pipeline through Iran in defiance of U.S. policy.

Pelletiere, Stephen. *Iraq and the International Oil System: Why America Went to War in the Gulf.* New York: Praeger, 2001. Argues that the 1991 Gulf War was caused by a crisis in control of the world's oil supply. Pelletiere provides background on the rise of the international oil system from the 1920s and the rivalry between the giant multinational oil companies and OPEC in the 1970s.

Rashid, Ahmed. *Taliban: Islam, Oil and the New Great Game in Central Asia.* Rev. ed. New York: I. B. Tauris, 2002. Veteran Afghanistan reporter reviews the Taliban regime and its fall, including the role of Western nations and international companies, who are competing to build oil and gas pipelines through the country.

Salazar-Carillo, Jorge, and Bernadette West. *Oil and Development in Venezuela during the 20th Century.* New York: Praeger, 2004. Stresses the oil sector's contribution to the growth of the Venezuelan economy.

Sawyer, Suzanna. *Crude Chronicles: Indigenous Politics, Multinational Oil, and Neoliberalism in Ecuador.* Durham, N.C.: Duke University Press, 2004. Describes emergence of a strong indigenous movement during the 1990s and its struggles with both a multinational oil company and the Ecuadoran government because of alleged environmental and social damage caused by oil projects.

van der Leeuw, Charles. *Oil and Gas in the Caucusus and Caspian: A History.* New York: Palgrave Macmillan, 2000. Comprehensive overview of the origins and development of the oil and gas industry in this key area.

van der Linde, Cobe. *The State and the International Oil Market: Competition and the Changing Ownership of Crude Oil Assets.* New York: Kluwer Academic Publishers, 2000. Claims that the competitive position of the multinational oil companies has improved relative to that of the oil-producing countries since the 1970s because the former have lowered their costs, while the costs of the latter have risen due to organizational efficiencies.

Wakeman-Linn, John. *Managing Oil Wealth: The Case of Azerbaijan.* Ogdensburg, N.Y.: Renouf Publishing Co., Ltd., 2004. Describes the effects of large oil revenues on Azerbaijan, a former member of the Soviet Union on the shore of the Caspian Sea.

Wang, Haijiang Henry. *China's Oil Industry and Market.* St. Louis, Mo.: Elsevier, 1999. Analyzes China's attempts to meet its soaring demand for oil by domestic production and its efforts to pay for oil imports and to make necessary revisions to its policies.

Energy Supply

ARTICLES

Adelman, M. A. "The Real Oil Problem." *Regulation*, vol. 27, Spring 2004, pp. 16–21. Adelman claims that world oil problems today are due to OPEC, not to oil shortages.

Al-Chalabi, Issam. "Iraqi Oil Policy: Present and Future Perspectives." *Oil and Gas Journal*, vol. 102, March 22, 2004, pp. 42–48. The three major issues facing Iraq in regard to its oil industry are its huge oil potential, domestic politics, and future oil policy.

Aris, Ben. "Putin Returns to State Control to Go Forward." *Euromoney*, vol. 35, January 2004, pp. 130–133. Whatever he may say, Russian president Vladimir Putin is prosecuting oil magnate Mikhael Khodorkovsky at least partly as part of a plan to regain state control of the Russian oil industry.

Ayres, Ed. "'It's Not About Oil!'" *World Watch*, vol. 16, May–June 2003, pp. 3–4. Many wars and conflicts around the world in the past half century have been caused or exacerbated by attempts to control oil resources.

Badcock, James. "Western Firms Scramble for Contracts." *African Business*, June 2004, pp. 58–59. Western oil companies are flocking back to Libya now that the United States has lifted its economic sanctions against that country.

Barlett, Donald L., and James B. Steele. "Iraq's Crude Awakening." *Time*, vol. 161, May 19, 2003, pp. 49–50. The future of Iraq's oil production could have major effects, not only on the United States, but also on countries around the world.

———. "The Oily Americans." *Time*, vol. 161, May 19, 2003, p. 53. For half a century, the U.S. government has secretly meddled in Middle Eastern politics, especially in Iran and Afghanistan, for oil-related reasons.

"Big Oil's Dirty Secrets." *The Economist*, vol. 367, May 10, 2003, n.p. Lawsuits are highlighting alleged unethical practices of multinational oil corporations around the world.

Brown, Frank, and Nadezhda Titova. "The Yukos Endgame." *Newsweek International*, July 5, 2004, p. 47. Russian President Vladimir Putin's arrest and trial of Mikhael Khodorkovsky, former CEO of the Yukos oil company, may have many motives, but one of them is almost surely effective renationalization of the company.

Bruno, Kenny. "De-Globalizing Justice." *Multinational Monitor*, vol. 24, March 2003, pp. 13–16. Human rights organizations are using the Alien Tort Claims Act to sue American-based transnational corporations, including oil giants such as ChevronTexaco, for complicity in alleged abuses carried out by governments in countries where the multinationals do business; the businesses have responded by attacking the law.

Annotated Bibliography

Caesar, Mike. "Maracaibo's Pipeline of Wealth and Woes." *Americas*, vol. 55, March–April 2003, pp. 6–13. Maracaibo, Venezuela's second-largest city, has both gained and suffered from the area's oil wealth.

"Can Oil Ever Help the Poor? Rags to Riches." *The Economist*, vol. 369, December 6, 2003, p. 39US. A watchdog group in the African nation of Chad is attempting to see that the country's government spends its new oil revenue wisely.

Cohen, Nick. "The Curse of Black Gold." *New Statesman*, vol. 132, June 2, 2003, pp. 25–27. In countries with widespread poverty and dictatorial governments, oil wealth almost always makes conditions grow worse for most of the population.

Collum, Randall, Jr., et al. "An In-Depth Analysis Shows Caspian's Oil Potential." *World Oil*, vol. 225, March 2004, pp. 69–75. The Caspian region will influence global energy markets and geopolitics for years to come.

Ebel, Robert. "Untapped Potential: The Future of Russia's Oil Industry." *Harvard International Review*, vol. 25, Spring 2003, pp. 26–31. The Russian oil industry has great potential, but investment from multinational corporations will be needed to improve its infrastructure.

Eifert, Benn, Alan Gelb, and Nils Borje Tallroth. "Managing Oil Wealth." *Finance & Development*, vol. 40, March 2003, pp. 40–44. Ability to manage a country's oil wealth depends on its type of government.

Eviatar, Daphne. "Africa's Oil Tycoons." *Nation*, vol. 278, April 12, 2004, pp. 11ff. Conditions worsen for Angola's poor, but oil wealth makes life good for the country's elite.

———, and Christopher Dickey. "What Oil Wants." *Newsweek International*, March 24, 2003, p. 38. Oil companies favor contracts with governments called production-sharing agreements, but smart oil-producing countries avoid this type of contract.

Eytchison, Patrick. "The Caspian Oil Myth." *Synthesis/Regeneration*, Fall 2003, p. 48. Eytchison claims that the Caspian Sea area's oil and gas potential is far less than the U.S. Geological Service and Energy Information Administration have stated.

"Firms Partner Locally for Sustained Community Development." *World Oil*, vol. 224, March 2003, pp. S28–S30. Multinational oil-related firms such as ChevronTexaco sponsor educational and other aid and development programs in Angola.

Fischer, Perry A. "Sustainable Development Efforts Succeed in Amazon Jungle Project." *World Oil*, vol. 224, December 2003, pp. 50–52. Claims that Occidental Petroleum's Block 15 development and pipeline in Ecuador are being handled in a way that will benefit the company, the country, and local residents.

Fisher, Daniel. "Dangerous Liaisons." *Forbes*, vol. 171, April 28, 2003, pp. 84 ff. International companies such as ExxonMobil are willing to make deals with dictators in any country "where the oil is."

Ford, Neil. "Oil: Ethics vs. Profits." *African Business*, November 2000, p. 26. The use of oil revenues by the governments of Angola and some other African countries has been criticized, but it is unclear how much blame for government corruption and other abuses multinational oil companies should shoulder.

Fox, Justin. "OPEC Has a Brand-New Groove." *Fortune*, vol. 146, October 14, 2002, pp. 115–116. OPEC faces a difficult balancing act in trying to set optimum oil prices, and many factors that affect international oil prices are beyond its control.

Freedman, Michael. "Supping with the Devil." *Forbes Global*, vol. 6, April 28, 2003, p. 34. The suit against California oil company Unocal for alleged complicity in government use of forced labor in Burma (Myanmar) is one of several suits brought by labor rights groups against multinational companies under the Alien Tort Claims Act.

Fri, Robert W. "U.S. Oil Dependence Remains a Problem." *Issues in Science and Technology*, vol. 19, Summer 2003, pp. 53–54. OPEC is likely to continue to control oil prices for the foreseeable future; the only way for the United States to escape that control is to reduce its oil consumption.

Harvan, Rob. "Yukos: Evolution into a Global Company." *World Refining*, vol. 14, January–February 2004, pp. 30–33. Describes the development of Yukos, and the Russian oil industry in general, from its terrible condition at the time of privatization in the early 1990s to a modern, competitive business.

Hepburn, Donald F. "Is It a War for Oil?" *Middle East Policy*, vol. 10, Spring 2003, pp. 29–34. American-based international oil companies will not benefit substantially from a U.S. takeover of Iraq; the country's oil revenues will be needed for internal rebuilding.

"How Did They Do It?" *Latin Finance*, November 2003, pp. 50–51. Explains how Venezuela made a successful comeback in international oil markets after its disastrous national strike in December 2002.

"In the Pipeline: The Oil Wars." *The Economist*, vol. 371, May 1, 2004, p. 1. Discusses disputes among Asian countries over oil.

Jermyn, Leslie. "Fueling Disaster." *This Magazine*, vol. 37, September–October 2003, pp. 12–15. Local citizens and environmental groups have protested the building of a pipeline carrying oil from the Amazon area in Ecuador, but oil companies say their work on it is ethical.

Jobin, William. "Health and Equity Impacts of a Large Oil Project in Africa." *Bulletin of the World Health Organization*, vol. 81, June 2003, pp. 420–426. The World Bank Group appointed an international panel of experts to make recommendations for minimizing adverse health and environmental impacts of the Chad-Cameroon Pipeline Project, but the experts' advice was largely ignored, a member of the panel says.

Klare, Michael T. "Oil Moves the War Machine." *The Progressive*, vol. 66, June 2002, pp. 18–19. The war on terrorism and the struggle for oil have become one vast global enterprise.

Kleveman, Lutz. "Oil and the New 'Great Game.'" *The Nation*, vol. 278, February 16, 2004, pp. 11–12. By supporting dictators in order to protect oil interests in the Caspian Sea region, the United States is increasing recruitment for anti-American terrorist groups.

Kretzmann, Steve. "Oil, Security, War: The Geopolitics of U.S. Energy Planning." *Multinational Monitor*, January–February 2003, pp. 11–16. The George W. Bush administration sees protecting U.S. access to global oil supplies as essential to national security, but a better approach to security would be to reduce the country's dependence on oil.

Lavelle, Marianne. "Russian Rigs to the Rescue." *U.S. News & World Report*, February 11, 2002, p. 48. The newly powerful Russian oil industry is helping to lower world oil prices, but foreigners are still leery of investing in the country.

Liu, Melinda. "Hungry for Power." *Newsweek International*, May 3, 2004, p. 38. Chinese officials are scouring the globe for the energy they need to fuel their growing economy.

"The Long Slump." *Global Markets*, vol. 33, February 10, 2003, p. 1. A focus on national self-interest is replacing the globalism of the 1990s in the oil business and will lead to an economic slump.

Losman, Donald. "Oil Is Not a National Security Issue." *USA Today*, vol. 130, January 2002, pp. 16–20. The economic damage attributed to the Arab 1973–74 oil embargo was actually caused by poor policy decisions in the United States, but fear of the "oil weapon" has shaped U.S. foreign policy ever since. Real or perceived oil shortages should be dealt with by economic, not military, means.

Macewan, Arthur. "Is It Oil?" *Dollars and Sense*, May–June 2003, pp. 20–24. Control of Iraq's oil resources is not the only reason for the U.S. war against that country, but it is a major factor.

Mandil, Claude. "A Steadying Influence in Uncertain Times." *Petroleum Economist*, vol. 70, August 2003, pp. 15–17. Interview with Claude Mandil, new director of the International Energy Agency, discusses the agency's policies for averting oil shortages and using strategic oil reserves, as well as its relationships with OPEC and the European Union.

Miller, Jim. "Sustainable Development." *Oil and Gas Investor*, vol. 23, August 2003, pp. 44–46. Multinational oil companies will need to make sustainable development a priority if they wish to stay profitable.

Naim, Moises. "Russia's Oily Future." *Foreign Policy*, January–February 2004, pp. 96–97. Russia's combination of oil wealth and weak democracy threaten to make it a "petro-state," heir to many political and economic ills.

Nivola, Pietro S. "Energy Independence or Interdependence?" *Brookings Review*, vol. 20, Spring 2002, pp. 24–27. The United States should buy more of its energy sources from Canada and Mexico, and laws in all three countries should be changed to encourage such activity.

Perz, Nathan. "The Future of War for Oil: The Caspian Basin and Shifting U.S. Policy in Central Asia." *Synthesis/Regeneration*, Spring 2003, pp. 41–44. The Caspian Sea area could become "a new Middle East" in terms of both oil resources and political tension.

"A Positive Force." *Petroleum Economist*, vol. 69, August 2002, p. 2. Enlightened self-interest is forcing oil companies to take more responsibility for the environmental and social effects of their operations in developing countries.

Powell, Bill. "Russia Pumps It Up." *Fortune*, vol. 145, May 13, 2002, pp. 84–85. The Russian oil industry is booming and could provide a welcome alternative to OPEC as a source of supply for the United States.

Reed, Stanley. "Oil Shortage? Saudi Arabia." *Business Week*, April 5, 2004, pp. 28 ff. There is plenty of oil still in the ground of this key Middle Eastern country, but it will not be easy to obtain. Multinational oil companies hope that the Saudi government will seek help from them.

———. "Waiting for the Oil to Flow Again." *Business Week*, June 21, 2004, p. 42. Iraq's oil prospects seemed favorable at the time of Saddam Hussein's overthrow, but a year later they looked dimmer because of unrelenting sabotage. Return of control to an Iraqi government may improve them.

Reid, Keith. "Oil Power." *National Petroleum News*, vol. 96, May 2004, pp. 20–21. Reid describes the contributions of the U.S. petroleum industry to Allied victories in both world wars.

"Reliance on Oil Income Challenges Those on Road to Democracy." *The Oil Daily*, vol. 53, July 14, 2003, n.p. Oil wealth and democracy can coexist in a country, but this usually happens only when democratic traditions are in place before oil becomes a major economic factor.

"Roads to Freedom." *Global Markets*, vol. 33, May 19, 2003, p. 1. The basic economics of oil seem to be antithetical to free markets and open societies.

"Saudi Terror Attack Aimed at Oil Industry Ratchets up Pressure." *Oil Daily*, vol. 54, June 2, 2004, n.p. Foreign oil companies in Saudi Arabia are reviewing security procedures and even withdrawing staff in the wake of a terrorist attack on a resort that left more than 22 people dead, including 19 foreigners.

Searight, Sarah. "Region of Eternal Fire." *History Today*, vol. 50, August 2000, pp. 45 ff. History of oil development around Baku, on the southwestern shore of the Caspian Sea, beginning in the 1870s.

Smith, Pamela Ann. "Bush and Blair Scramble for Oil." *The Middle East*, August–September 2003, pp. 32–36. Claims that George W. Bush and Tony Blair invaded Iraq because of oil and may invade other countries to secure future oil supplies as well.

Sperling, Daniel, and Eileen Clausen. "The Developing World's Motorization Challenge." *Issues in Science and Technology*, vol. 19, Fall 2002, pp. 59–66. Managing the rapid motorization of the developing world in an ecologically sound manner presents many challenges, but it can be done.

Surowiecki, James. "The Oil Weapon." *The New Yorker*, vol. 78, February 10, 2003, p. 38. Middle Eastern countries are not likely to repeat their 1973 oil embargo against the United States because doing so would hurt them more than it would hurt America.

Wallin, Tom, and David Knapp. "How Global Oil Markets Have Changed in 40 Years." *Oil Daily*, vol. 51, November 5, 2001, n.p. The global oil market's evolution toward free market mechanisms during the last four decades has had both good and bad effects on the industry.

"Watchdog or Lapdog?" *Global Markets*, vol. 33, September 15, 2003, p. 1. The International Energy Agency needs to protect consumers from the major oil companies, which use low stocks of oil to keep prices high and oppose release of strategic reserves.

Weissman, Robert. "Deadly Drilling in Aceh." *Multinational Monitor*, July 2001, p. 7. A lawsuit filed by the International Labor Rights Fund claims that ExxonMobil is responsible for human rights abuses committed by the Indonesian military during a company-sponsored oil drilling project in Aceh province, North Sumatra.

Wirth, Timothy E., C. Boyden Gray, and John D. Podesta. "The Future of Energy Policy." *Foreign Affairs*, vol. 82, July–August 2003, pp. 132 ff. U.S. energy policy has failed to address three great challenges: security risks arising from the world's dependence on oil; the risk to the global environment from climate change, caused primarily by burning of fossil fuels; and the lack of access of the world's poor to modern energy services and other basic tools for economic advancement.

Woolsey, James. "Defeating the Oil Weapon." *Commentary*, vol. 114, September 2002, pp. 29–34. The United States needs to take steps, including increased energy efficiency, to weaken the power of Saudi Arabia's "swing capacity" in the world oil market.

Zagorin, Adam. "Slave Labor?" *Time*, vol. 162, November 24, 2003, p. A18. The International Labor Rights Fund is suing California oil company

Unocal for complicity in the Burmese (Myanmar) government's alleged use of forced labor on a pipeline project.

Web Documents

Baue, William. "Does Supreme Court Validation of Alien Tort Claims Act Apply to Corporations?" Corporate Social Responsibility Newswire Service. Available online. URL: http://www.csrwire.com/sfarticle.cgi?id=1458. Posted July 1, 2004. The Supreme Court decision in *Sosa v. Alvarez-Machain* discusses the applicability of the Alien Tort Claims Act to overseas human rights violations but does not make clear whether ATCA can be used against oil companies allegedly complicit in such violations on oil projects in developing countries.

"Behind the Shine: The Other Shell Report." Refinery Reform Campaign. Available online. URL: http://www.refineryreform.org/downloads/shellreport_behindtheshine.pdf. Posted in June 2004. Report accuses multinational oil giant Shell of causing severe environmental and other damage on its projects, citing examples from around the world.

"Bus Systems for the Future: Achieving Sustainable Transport Worldwide." International Energy Agency. Available online. URL: http://www.iea.org/dbtw-wpd/textbase/nppdf/stud/02/bussystems.pdf. Posted in 2002. Provides examples of bus systems that are bringing clean fuels and sustainable transport to developing countries, with a focus on the bus rapid transit systems emerging in Latin America.

"Caspian Sea Region." Energy Information Administration. Available online. URL: http://www.eia.doe.gov/emeu/cabs/caspian.html. Posted in December 2004. EIA background report on this key oil-producing area, consisting of Russia, Iran, and three former Soviet Union countries.

"China's Worldwide Quest for Energy Security." International Energy Agency. Available online. URL: http://www.iea.org/dbtw-wpd/textbase/nppdf/free/2000/china2000.pdf. Posted in 2000. China's rapid economic growth is sparking a surging demand for energy, much of which will have to be imported. The country is therefore using trade and investment to try to gain a more prominent place in the global energy market.

"*Doe v. Unocal.*" EarthRights International. Available online. URL: http://www.earthrights.org/unocal/index.shtml. Last updated October 6, 2003. Describes group of lawsuits growing out of the California-based oil company's alleged complicity in human rights violations committed by the military of Myanmar (Burma) during construction of a gas pipeline in which Unocal was an investor. The web page includes links to court decisions and related documents. EarthRights International is one of the sponsors of the suits.

"The Energy Tug of War: The Winners and Losers of World Bank Fossil Fuel Finance." Sustainable Energy and Economy Network. Available online. URL: http://www.seen.org/PDFs/Tug_of_war.pdf. Posted in April 2004. Claims that World Bank fossil fuel projects benefit northern corporations such as Halliburton, not developing countries in the south.

"Engaging the Private Sector in the Clean Development Mechanism." World Business Council for Sustainable Development. Available online. URL: http://www.wbcsd.ch/DocRoot/0sDGYastHe3029UiRnVO/cdm.pdf. Posted in March 2004. Suggests options for increasing private sector engagement in development of low-carbon-emission energy in developing countries called for by the Kyoto Protocol. The web page reviews previous CDM projects.

"Foreign Corrupt Practices Act: Antibribery Provisions." U.S. Department of Justice. Available online. URL: http://www.usdoj.gov/criminal/fraud/fcpa/dojdocb.htm. Downloaded on July 2, 2004. Describes this law, passed in 1977, which officials of some oil companies have been accused of violating.

"Fueling Poverty: Oil, War, and Corruption." Christian Aid. Available online. URL: http://www.christianaid.org.uk/indepth/0305cawreport/cawreport03.pdf. Posted in May 2003. Explains how oil wealth has "fueled poverty," government corruption and repression, and environmental damage in Angola, Sudan, and Kazakhstan.

Gary, Ian. "Bottom of the Barrel: Africa's Oil Boom and the Poor." Catholic Relief Services. Available online. URL: http://www.catholicrelief.org/get_involved/advocacy/policy_and_strategic_issues/oil_report_full.pdf. Posted in June 2003. Asserts that money from the sale of oil resources has increased corruption and poverty in many countries in Africa, Asia, and Latin America.

"Going Where the Oil Is." Petroleum Industry Research Foundation, Inc. Available online. URL: http://www.pirinc.org/download/goingwheretheoilis.pdf. Posted in July 2003. Describes the challenges that multinational oil companies face because most of the world's oil is located in countries with undemocratic governments, and the companies must deal with those governments in order to obtain the oil. The web page discusses corporate responsibility for government behavior during projects.

"*International Energy Annual* Overview 2002." Energy Information Administration. Available online. URL: http://www.eia.doe.gov/emeu/iea/overview.html. Last modified on October, 4, 2004. Covers world primary energy production trends, major energy producers and consumers, regional energy production and consumption, statistics for specific fuels, and carbon dioxide emissions from burning of fossil fuels.

Martinez, Nadia. "Destabilizing Investments in the Americas II: The Inter-American Development Bank's Fossil Fuel Financing 1992–2004." Sustainable Energy and Economy Network. Available online. URL: http://www.seen.org/PDFs/energy_report_final.pdf. Posted in March 2004. Alleges that the IDB is using billions in taxpayer money to support destructive fossil fuel projects throughout Latin America and the Caribbean and does not consider the climate impacts of its lending.

Mitchell, John V. "Renewing Energy Security." Royal Institute of International Affairs. Available online. URL: http://www.riia.org/pdf/research/sdp/Renewing%20Energy%20Security%20Mitchell%20July%202002.pdf. Posted on July 20, 2002. Considers energy security issues, including climate change and resource scarcity, in developed and developing countries in the short, medium, and long term. Mitchell concludes that international trade and investment provide the best route both to energy security and to protecting the environment.

"Mobility 2001: World Mobility at the End of the Twentieth Century and Its Sustainability." World Business Council for Sustainable Development. Available online. URL: http://www.wbcsd.org/web/projects/mobility/english_full_report.pdf. Posted in 2001. Discusses patterns of mobility in the developed and developing world and the challenges to sustainability that rapid growth in mobility presents, especially in the cities of developing countries.

"Mobility 2030: Meeting the Challenges to Sustainability." World Business Council for Sustainable Development. Available online. URL: http://www.wbcsd.ch/DocRoot/nfGx0tY87RjAQlIxqLAt/mobility-full.pdf. Posted in July 2004. Final report of the WBCSD's Sustainable Mobility Project offers a vision of global road transportation of people, goods, and services; identifies seven sustainable mobility goals; and establishes a set of indicators to help measure the effectiveness of the various options.

"North Sea." Energy Information Administration. Available online. URL: http://www.eia.doe.gov/emeu/cabs/northsea.html. Posted in August 2004. Background report on this region, mostly co-owned by Norway and Great Britain, which contains Europe's largest oil and natural gas reserves and is a leading non-OPEC fossil-fuel-producing area.

"OPEC General Information." Organization of Petroleum Exporting Countries. Available online. URL: http://www.opec.org/publications/gi/geninfo.pdf. Downloaded on July 23, 2004. General information about this influential organization, including its founding, principal aims, and structure.

Ross, J. P. "Non-Grid Renewable Energy Policies: International Case Studies." Center for Resource Solutions. Available online. URL: http://www.resource-solutions.org/Library/librarypdfs/DistRural-

Non-gridREExperiences.pdf. Posted on August 16, 2001. Provides a thorough review of federal and local governmental policies that can be helpful in promoting rural, off-grid renewable energy development.

"Russia." Energy Information Administration. Available online. URL: http:///www.eia.doe.gov/emeu/cabs/russia.pdf. Posted in May 2004. Country analysis report on this major holder of fossil fuel reserves, fossil fuel exporter, and energy consumer.

"The Story You Haven't Heard About . . . the Activists' Lawsuits." Unocal. Available online. URL: http://www.unocal.com/myanmar/suit.htm. Downloaded on July 2, 2004. This and related pages on the Unocal site give the oil company's response to the lawsuits alleging its complicity in human rights violations committed by Myanmar (Burma) military forces during construction of the Yadana pipeline, a project in which Unocal was an investor.

"Toward an International Energy Trade and Development Strategy." U.S. Energy Association. Available online. URL: http://www.usea.org/T&Dreport.pdf. Posted in October 2001. Discusses energy trade among the United States, Canada, and Mexico; energy trade sanctions; energy as an economic driver for international development; and promotion of international energy trade.

Vallette, Jim, Steve Kretzmann, and Daphne Wysham. "Crude Vision: How Oil Interests Obscured U.S. Focus on Chemical Weapons Use by Saddam Hussein." Sustainable Energy and Economy Network Institute for Policy Studies. Available online. URL: http://www.seen.org/PDFs/Crude_Vision2.pdf. Posted on August 13, 2002. Claims that in the 1980s, the Reagan administration ignored evidence that Iraq was using chemical weapons while it negotiated with Saddam Hussein to obtain access to Iraqi oil.

"Venezuela." Energy Information Administration. Available online. URL: http://www.eia.doe.gov/emeu/cabs/venez.html. Posted in June 2004. Country analysis background report on this leading South American oil exporter, the only non-Arab charter member of OPEC.

"World Energy 'Areas to Watch.'" Energy Information Administration. Available online. URL: http://www.eia.doe.gov/emeu/cabs/hot.html. Posted in August 2004. Briefly describes countries or regions confronting significant political, economic, or other issues that could affect domestic or world oil and gas markets.

"World Energy Outlook 2002: Energy and Poverty." International Energy Agency. Available online. URL: http://www.worldenergyoutlook.org/weo/pubs/weo2002/energypoverty.pdf. Posted in 2002. Indicates that a reliance on the most primitive fuels—wood, agricultural waste, and animal excrement—is a mark of poverty worldwide and that alleviating that poverty requires access to electricity. Electricity will usually need to come from fossil fuels and conventional grids rather than renewable projects.

ENVIRONMENTAL ISSUES

BOOKS

Baumert, Kevin A., ed. *Building on the Kyoto Protocol: Options for Protecting the Climate.* Washington, D.C.: World Resources Institute, 2002. Seventeen contributors from nine countries explore options in addition to the Kyoto Protocol for promoting long-term climate protection.

Beamish, Thomas D. *Silent Spill: The Organization of an Industrial Crisis.* Cambridge, Mass.: MIT Press, 2002. Describes an oil spill—probably the nation's largest—that persisted unattended for 38 years in California's Guadelupe Dunes as an example of environmental "blind spots" that are ignored because they do not cause social disruption or dramatic visible destruction.

Chandler, William. *Energy and Environmental Policies in the Transition Economies.* Boulder, Colo.: Westview Press, 2000. Examines environmental politics in the former Soviet states and Eastern Europe, including their potential for curbing greenhouse gas emissions and the reforms that will be necessary to make this happen.

Elliott, David. *Energy, Society and the Environment: Technology for a Sustainable Future.* 2nd ed. New York: Routledge, 2003. Claims that social, political, and economic as well as technological solutions for energy-related environmental problems are likely to be necessary.

Eno Transportation Foundation. *Global Climate Change and Transportation: Coming to Terms.* Washington, D.C.: Eno Transportation Foundation, 2002. Presents views of the relationship between transportation, especially in the United States, and global climate change.

Estrada, Javier, Kristian Tangen, and Helge Ole Bergeson. *Environmental Challenges Confronting the Oil Industry.* Hoboken, N.J.: John Wiley & Sons, 1999. Studies strategies of oil companies, refineries, and industry organizations for responding to environmental challenges, especially those related to climate change.

Gelbspan, Ross. *The Heat Is On: The Climate Crisis, the Cover-up, the Prescription.* Rev. ed. Cambridge, Mass.: Perseus Press, 1998. Claims that oil and coal companies and conservative politicians have conspired to undermine public faith in science and thereby defer action against global warming.

Hinrichs, Roger A., and Merlin Kleinbach. *Energy: Its Use and the Environment.* 3rd ed. Pacific Grove, Calif.: Brooks/Cole, 2001. Introductory college textbook emphasizes physical principles that underlie energy use and its environmental effects.

Annotated Bibliography

Khan, Namir, and Willem H. Vanderburg. *Sustainable Energy: An Annotated Bibliography.* Lanham, Md.: Rowman & Littlefield, 2001. Introduces readers to approaches for making the energy requirements of modern life less environmentally damaging and more sustainable.

Kursonoglu, Behram N., Stephan L. Mintz, and Arnold Perlmutter, eds. *Global Warming and Energy Policy.* New York: Plenum Press, 2001. Proceedings of a symposium on the subject explore two issues: the possibility that Earth will become warmer because of human-caused carbon dioxide emissions, and the health effects of particulates emitted by fossil fuel combustion.

Meyer, Aubrey. *Contraction and Convergence: The Global Solution to Climate Change.* Dartington, Totnes, Devon, U.K.: Green Books, 2000. Proposes that greenhouse gases first be reduced, especially by developed countries, and then reduction quotas be assigned on a per capita basis.

National Research Council. *Cumulative Environmental Effects of Oil and Gas Activities on Alaska's North Slope.* Washington, D.C.: National Academies Press, 2003. Efforts by the oil industry and regulatory agencies have reduced many damaging environmental effects, but such effects have not been completely eliminated.

———. *Implementing Climate and Global Change Research: A Review of the Final U.S. Climate Change Science Program Strategic Plan.* Washington, D.C.: National Academies Press, 2004. Reviews a draft strategic plan from the U.S. Climate Change Science Program, formed in 2002 to direct and coordinate U.S. research efforts in this field.

Ocean Studies Board and Marine Board. *Oil in the Sea III: Inputs, Fates, and Effects.* Washington, D.C.: National Academies Press, 2003. Studies input of oil into the sea, its fate, and its biological effects. Appendices give statistics on the amount of oil leaked into the sea from various human and natural sources.

Park, Patricia D. *Energy Law and the Environment.* Washington, D.C.: Taylor & Francis, 2001. First book to consider relationships between energy law and environmental law as they affect the industrial sector.

Skjaerseth, John Birger, and Tora Skodvin. *Climate Change and the Oil Industry: Common Problems, Different Solutions.* Manchester, U.K.: Manchester University Press, 2004. Extensive study of the climate strategies of ExxonMobil, Shell, and Statoil generates information about whether and how corporate resistance to policies to prevent climate change can be overcome.

Smith, Eric R. A. N. *Energy, the Environment, and Public Opinion.* Lanham, Md.: Rowman & Littlefield, 2002. Investigates what shapes public opinion on energy and environmental issues such as oil development and nuclear power.

Trustees for Alaska. *Under the Influence: Oil and Industrialization in America's Arctic.* Anchorage: Trustees for Alaska, 1998. Updated report of pollution problems and other environmental impacts caused by oil industry activities in the Arctic. The book alleges that the industry has unfairly influenced public policy about development in Alaska.

Watts, Robert G., ed. *Innovative Energy Strategies for CO_2 Stabilization.* New York: Cambridge University Press, 2002. Ten papers evaluate increased energy efficiency, renewable energy sources, and atomic fission and fusion as ways of minimizing CO_2 emissions.

World Energy Council. *Living in One World.* London: World Energy Council, 2001. Studies links between human activities, including those related to energy, and the environment and describes positive and negative future scenarios for the planet.

Wunder, Sven. *Oil, Wealth and the Fate of the Forest.* New York: Routledge, 2003. Uses case studies based on eight countries to show how oil revenues can indirectly come to protect tropical rain forests.

ARTICLES

Baird, Vanessa. "The Big Switch." *New Internationalist,* June 2003, pp. 9–12. Global warming is already happening, and a rapid switch to renewable, noncarbon energy sources is essential, but governments and large energy corporations are resisting such a change.

Davis, Devra. "Fossil Fuel's Fear Factor." *USA Today,* vol. 132, January 2004, pp. 28–29. One way to discourage society's addiction to fossil fuels is to make their price reflect the costs of their damage to the environment and human health.

Drummond, Steve. "Coming up for Air." *Power Economics,* vol. 7, March 2003, pp. 22–24. Describes progress in Britain and the European Union in meeting Kyoto Protocol goals, stressing the importance of emissions trading schemes.

Frum, David. "The Air War at Home." *National Review,* vol. 54, December 23, 2002, n.p. The New Source Review policy, praised by environmentalists, has hidden costs, and the George W. Bush administration's discontinuance of it is wise.

Gardiner, Stephen M. "The Global Warming Tragedy and the Dangerous Illusion of the Kyoto Protocol." *Ethics & International Affairs,* vol. 18, April 2004, pp. 23–40. Claims that the Kyoto Protocol, even if implemented, is too weak to slow global warming and protect the environment for future generations.

Hamilton, Kirsty. "Atmospheric Pressure." *Petroleum Economist,* vol. 70, December 2003, p. 32. Pressure on oil and gas companies to comply with laws aimed at preventing climate change is growing, and their business success may come to depend on their willingness to respond.

Harrison, Gareth, and Bert Whittington. "Climate Change—A Drying up of Hydropower Investment?" *Power Economics*, vol. 6, January 2002, pp. 25–27. By reducing river flow and causing other changes in the availability of water, global warming could reduce the efficiency of electricity generation from hydropower plants, thus discouraging continued investment in such plants.

Hertsgaard, Mark. "A New Ice Age?" *The Nation*, vol. 278, March 1, 2004, p. 9. Report by elite Pentagon planning unit declares that climate change is a major national security threat.

———. "Trashing the Environment." *The Nation*, vol. 276, February 3, 2003, pp. 15–16. Summarizes the George W. Bush administration's alleged attacks on environmental legislation during its first two years in office.

Hillman, Mayer. "The Options Are Awesome." *Energy and Environmental Management*, January–February 2003, pp. 16–17. Substantial social changes, including limits to economic growth, will be necessary to avoid the effects of global warming.

Hirsch, Robert L. "Ammunition for the Energy Wars." *Oil and Gas Investor*, vol. 23, September 2003, p. 10–14. Offers questions for policymakers concerning three key environmental issues: renewable energy, the hydrogen economy, and global warming.

Hoffert, Martin I., et al. "Advanced Technology Paths to Global Climate Stability." *Science*, vol. 298, November 1, 2002, pp. 981–987. Surveys technological approaches to stabilizing global climate while allowing economic development and concludes that all have severe deficiencies; intensive research is urgently needed.

Parker, Larry B., and John E. Blodgett. "Electricity Restructuring and Air Quality." *Forum for Applied Research and Public Policy*, vol. 16, Fall 2001, pp. 70–76. Electricity deregulation could lead to increased air pollution from power plants if market conditions favor keeping older, coal-burning plants in operation.

"People Power." *New Internationalist*, June 2003, pp. 22–23. Around the world, grassroots environmental groups are making progress in their fight against global warming and the big corporations and fossil fuel use that contribute to it.

"Say Watt? Energy and the Environment." *The Economist*, vol. 366, March 1, 2003, n.p. Summarizes a new British government proposal for reducing greenhouse gas emissions.

Shelby, Ashley. "Whatever It Takes." *The Nation*, vol. 278, April 5, 2004, pp. 16 ff. Describes alleged legal maneuvering that Exxon used to avoid paying damages for the *Exxon Valdez* oil spill.

Stewart, Richard B., and Jonathan Wiener. "Practical Climate Change Policy." *Issues in Science and Technology*, vol. 20, Winter 2004, pp. 71–78. Proposes an alternative to the Kyoto Protocol, a worldwide cap-and-trade system involving all greenhouse gases, and cites advantages for U.S. participation in such an agreement.

Stone, Richard. "Caspian Ecology Teeters on the Brink." *Science*, vol. 295, January 18, 2002, pp. 430–432. The Caspian Sea's two valuable resources are caviar (sturgeon eggs) and oil—and harvesting both presents environmental dangers.

Thorning, Margo. "Heroes and Villains?" *Power Economics*, vol. 7, April 2003, pp. 26–27. Lists reasons for differences between the strategies of the United States and the European Union in dealing with the threat of global warming and claims that the U.S. strategy is based on more accurate assumptions.

"Warming Relations." *Weekly Petroleum Argus*, vol. 33, August 11, 2003, p. 1. Oil companies and environmentalists are less hostile to each other on the subject to global warming than they used to be, but their viewpoints and goals remain fundamentally different.

Wasserman, Harvey. "Power Struggle." *Multinational Monitor*, vol. 22, June 2001, pp. 9 ff. Claims that the state deregulation plan that helped to bring about the 2000–01 California energy crisis, as well as the federal policies of the George W. Bush administration, are attacks on environmental legislation and goals.

Zinsmeister, Karl. "Beware of Green Fairy Tales." *The American Enterprise*, vol. 12, September 2001, pp. 4 ff. Environmentalists unfairly demonize energy production and consumption.

Zuckerman, Mortimer B. "The Caribou Conundrum." *U.S. News & World Report*, vol. 130, April 30, 2001, p. 72. Drilling for oil in the Arctic National Wildlife Refuge probably will not cause much environmental damage.

WEB DOCUMENTS

"*Aguinda v. Texaco:* Questions and Answers." Texacorainforest.com. Available online. URL: http://www.texacorainforest.com/why/questions.html. Downloaded on July 2, 2004. Explains the suit by Ecuadoran residents and environmentalists against ChevronTexaco, claiming social and environmental damage during Texaco-sponsored drilling activities in Ecuador, from the plaintiffs' point of view.

American Institute of Petroleum. "Oceans and Oil Spills." World Petroleum Congress web site. Available online. URL: http://www.world-petroleum.org/education/ocean/index.html. Downloaded on July 23, 2004. Claims that most oil in the sea is natural rather than coming from the oil industry. The web page describes methods used to prevent spills and leaks and ways of cleaning up spills that occur.

Baue, William. "ChevronTexaco Faces Class-Action Suit in Ecuador over Environmental Damage." Socialfunds.com. Available online. URL: http://www.socialfunds.com/news/article.cgi/article1419.htm. Posted on May 11, 2004. Describes beginning of trial in Ecuador in suit of Ecuadoran residents against Texaco for alleged environmental damage during oil projects.

"Beyond Kyoto: Energy Dynamics and Climate Stabilisation." International Energy Agency. Available online. URL: http://www.iea.org/dbtw-wpd/Textbase/envissu/cop9/files/beyond_kyoto.pdf. Posted in 2002. Describes the energy sector's options for attempting to reduce climate change in an atmosphere of uncertainty and global inequity, including various types of international agreement.

"Climate Change 2001." Intergovernmental Panel on Climate Change. Available online. URL: http://www.grida.no/climate/ipcc_tar/. Downloaded on July 22, 2004. This report consists of four volumes: The Scientific Basis; Impacts, Adaptation, and Vulnerability; Mitigation; and Synthesis Report.

"Dams and Development." World Commission on Dams. Available online. URL: http://www.dams.org/report/contents.htm. Posted on November 16, 2000. Global review of economic, social, and environmental impacts of large hydropower dams. The page makes recommendations for lessening harmful effects.

Dobriansky, Paula J. "Addressing the Challenge of Global Climate Change." U.S. Global Climate Change Research Information Office. Available online. URL: http://www.gcrio.org/OnLnDoc/pdf/dobriansky031119.pdf. Posted on November 19, 2003. U.S. Undersecretary of State for Global Affairs expresses the belief that mitigation of climate change should be pursued as part of a broad strategic paradigm of sustainable development that balances environmental protection with the need for economic growth.

"Energy—the Changing Climate." Royal Commission on Environmental Pollution. Available online. URL: http://www.rcep.org.uk/newenergy.htm. Posted in June 2000. Discusses present and possible responses to climate change, with a focus on Britain, and makes recommendations for future British policy.

"Environmental Impacts of the Production of Electricity." European Small Hydropower Association. Available online. URL: http://www.esha.be/LCA_Study.pdf. Posted in July 2000. Comparative study of environmental impacts of eight technologies of electricity generation: lignite, coal, fuel oil, natural gas, nuclear, wind, small hydro, and solar photovoltaic.

Epstein, Paul R., and Jesse Selber, eds. "Oil: A Life Cycle Analysis of Its Health and Environmental Impacts." Harvard Medical School Center for

Health and the Global Environment. Available online. URL: http://www.med.harvard.edu/chge/fullreport.pdf. Posted in 2002. Describes the damage to human health and the environment resulting from each stage in the process of oil development and use: extraction, transport, refining, and combustion.

"ERCC White Paper on New Source Review." Electric Reliability Coordinating Council. Available online. URL: http://www.electricreliability.org/Statements/NSR-wp.htm. Downloaded on July 22, 2004. Claims that applying the Environmental Protection Agency's "new source review" antipollution requirements to what the group calls "routine maintenance" of power plants causes major problems, including increases in pollution, and that the market-based system established in the 1990 amendments to the Clean Air Act works better.

"External Costs: Research Results on Socio-Environmental Damages Due to Electricity and Transport." European Commission. Available online. URL: http://www.externe.info/externpr.pdf. Posted in 2003. Summary of results of the ExternE project, a 10-year study sponsored by the European Commission and the U.S. Department of Energy to determine the "external costs" of these energy uses, recommends that governments use taxation to force prices of polluting fuels to include these costs.

"Global Warming in Depth." Pew Center on Global Climate Change. Available online. URL: http://www.pewclimate.org/global-warming-in-depth/. Downloaded on June 16, 2004. Presents research from the Pew Center on economics, environmental impacts, international policy, and solutions for global warming.

"Greenhouse Gas Emission Trends and Projections in Europe: Final Draft." European Environment Agency. Available online. URL: http://reports.eea.eu.int/environmental_issue_report_2003_36/en. Posted on December 5, 2003. Assesses present and predicted future (2010) progress of the European Union and member countries toward reaching Kyoto Protocol emission targets and the likely effects of existing and proposed policies aimed at reducing emissions.

"Greenpeace Analysis of the Kyoto Protocol." Greenpeace. Available online. URL: http://archive.greenpeace.org/climate/politics/reports/kyoto.pdf. Posted on June 12, 1998. Detailed description of the Kyoto Protocol and critique of what the environmental organization regards as the agreement's weaknesses and loopholes.

Jorgenson, Dale W., et al. "U.S. Market Consequences of Global Climate Change." Pew Center on Global Climate Change. Available online. URL: http://www.pewclimate.org/docUploads/Market%5FConsequences%2Dreport%2Epdf. Posted in April 2004. Analyzes optimistic and pessimistic scenarios in terms of climate and economic effects.

Kruger, Joseph A., and William A. Pizer. "The EU Emissions Trading Directive: Opportunities and Potential Pitfalls." Resources for the Future. Available online. URL: http://www.rff.org/rff/Documents/RFF-DP-04-24.pdf. Posted in April 2004. Expresses concerns that uncertainty regarding aspects of the European Union's greenhouse gas emissions trading program will create volatile energy markets and compliance problems.

"MTBE at Center Stage." Petroleum Industry Research Foundation, Inc. Available online. URL: http://www.pirinc.org/download/mtbeatcenterstage.pdf. Posted in January 2004. Discusses the controversial oxygenating compound methyl tertiary butyl ether (MTBE), including issues of liability protection for refiners who used MTBE, the legal mandate that required oxygenating compounds to be added to gasoline, alternatives to MTBE, and the risks to gasoline supply/demand balances of trying to ban this compound.

"National Wildlife Refuges: Opportunities to Improve the Management and Oversight of Oil and Gas Activities on Federal Lands." United States General Accounting Office (U.S. Government Printing Office web site). Available online. URL: http://frwebgate.access.gpo.gov/cgi-bin/useftp.cgi?IPaddress=162.140.64.88&filename=d03517.txt&directory=/diskb/wais/data/gao.Report GAO=03=517. Posted on August 28, 2003. Report on oil and gas exploration and extraction on the 575 U.S. National Wildlife Refuges concludes that although there has been negligible harm in some cases, in others there was damage to wildlife and ecology from spills and other causes.

"New Source Review." American Petroleum Institute. Available online. URL: http://api-ec.api.org/policy/index.cfm?objectid=3E358364-86A4-11D5-BC6B00B0D0E15BFC&method=display_body&er=1&bitmask=001001003000000000. Downloaded on July 22, 2004. Position paper explains how and why this trade organization believes that the process of setting pollution emission standards for new and remodeled electric power plants needs to be modified for remodeled plants.

"Oil in the Arctic: Impacts of Oil Development on Alaska's North Slope." Trustees for Alaska. Available online. URL: http://www.trustees.org/. Downloaded on March 31, 2004. Series of fact sheets describes alleged environmental damage from air pollution, water pollution, waste disposal, water use, offshore drilling, spills, and more.

"Paybacks: Policies, Patrons and Personnel." Public Campaign. Available online. URL: http://www.publiccampaign.org/publications/studies/paybacks/Paybacks.pdf. Posted in September 2002. Claims that the George W. Bush administration is taking actions that damage the environment and awarding official posts to campaign contributors in oil and related industries.

"The Plain English Guide to the Clean Air Act." Environmental Protection Agency. Available online. URL: http://www.epa.gov/oar/oaqps/peg_caa/pegcaain.html. Downloaded on July 2, 2004. Describes chief features of the Clean Air Act as revised in 1990, the act's requirements for cleaning up pollution, pollution from mobile sources, and the acid rain program.

"Research and Current Activities." U.S. Climate Change Technology Program. Available online. URL: http://www.climatetechnology.gov/library/2003/currentactivities/car24nov03.pdf. Posted in November 2003. Summary of U.S. government-sponsored activities to limit greenhouse gas emissions and climate change includes reducing emissions from energy end-use and infrastructure, reducing emissions from energy supply, capturing and sequestering carbon dioxide, reducing emissions of other greenhouse gases, and enhancing capabilities to measure and monitor greenhouse gas emissions.

Schaeffer, Eric, and Abt Associates, Inc. "Particulate-Related Health Impacts of Eight Electric Utility Systems." Environmental Integrity. Available online. URL: http://64.78.32.98/pubs/PMimpacts8utilities.pdf. Posted in April 2002. Claims that pollution from 80 power plants constructed before 1973 was responsible for almost 6,000 premature deaths and tens of thousands of cases of illness from respiratory disease.

"Second API/OGP Conference on Voluntary Actions by the Oil and Gas Industry to Address Climate Change." American Petroleum Institute. Available online. URL: http://api-ec.api.org/filelibrary/APICCVAConfSYNOPSISFinalDraft2.pdf. Posted on November 21, 2002. Describes steps that the oil and gas industry is willing to take to limit climate change. Includes domestic (U.S.) and international expectations and technical discussions of such topics as carbon sequestration, alternative fuels, and carbon dioxide emissions reporting and trading.

"Smokestack Rollback." Refinery Reform Campaign. Available online. URL: http://www.refineryreform.org/downloads/Smokestack_Rollback.pdf. Posted in February 2002. Claims that changes to the Clean Air Act proposed by the George W. Bush administration, especially elimination of New Source Review for remodeled plants, will allow oil refineries to produce more pollution and jeopardize public health.

Sperling, Daniel, and Deborah Salon. "Transportation in Developing Countries: An Overview of Greenhouse Gas Reduction Strategies." Pew Center on Global Climate Change. Available online. URL: http://www.pewclimate.org/global-warming-in-depth/all_reports/transportation_overview/index.cfm. Posted in 2002. Discusses policy and strategy options for reducing greenhouse gas emissions from transportation in developing countries and makes recommendations.

Stavins, Robert N. "Can an Effective Global Climate Treaty Be Based on Sound Science, Rational Economics, and Pragmatic Politics?" Resources for the Future. Available online. URL: http://www.rff.org/rff/Documents/RFF-DP-04-28.pdf. Posted in May 2004. Proposes a three-part policy to substitute for the Kyoto Protocol.

"Technology Options for the Near and Long Term." U.S. Climate Change Technology Program. Available online. URL: http://www.climatetechnology.gov/library/2003/tech-options/tech-options.pdf. Posted in November 2003. A compendium of technology profiles and ongoing research and development at U.S. government agencies aimed at reducing greenhouse gas emissions in various sectors.

"Texaco in Ecuador." Texaco. Available online. URL: http://www.texaco.com/sitelets/ecuador/en. Downloaded on July 10, 2004. Texaco responds to accusations made by Ecuadoran and Peruvian residents and environmentalists in suits against the company for alleged environmental damage during a drilling project in Ecuador.

Vaughan, Scott, et al. "Environmental Challenges and Opportunities of the Evolving North American Electricity Market." Commission for Environmental Cooperation of North America. Available online. URL: http://www.cec.org/files/PDF//Electr-Vaughan_en.pdf. Posted in June 2002. Urges policy changes to increase electricity trading between Canada, the United States, and Mexico and addresses collective environmental concerns. Claims that it is possible to both meet electricity needs and protect health and environment if the three countries work together.

"A Viable Global Framework for Preventing Dangerous Climate Change." Climate Action Network. Available online. URL: http://www.climatenetwork.org/docs/CAN-DP_Framework.pdf. Downloaded on July 22, 2004. Describes predicted effects of different degrees of climate change and steps that countries should take to prevent or minimize the damage of such change, including setting of targets for greenhouse gas reduction.

CHAPTER 8

ORGANIZATIONS AND AGENCIES

Many organizations and groups, including government and intergovernmental agencies, trade and professional associations, and advocacy groups, deal with aspects of energy and related issues. The following entries include general-purpose organizations whose work concerns energy in whole or in part; organizations whose work relates to conventional fuels (fossil fuels and nuclear power); organizations whose work focuses on renewable energy sources and energy conservation; organizations whose work concerns electricity (as opposed to fuels from which electricity can be generated); and organizations whose work, in whole or in part, concerns environmental issues related to energy, such as global warming. Some of these organizations simply provide information on energy-related topics, while others take strong advocacy positions on one or more issues. Most organizations described in this chapter are in the United States, but some groups in Britain, Canada, and other countries are also listed.

In keeping with the widespread use of the Internet and e-mail, the web site address (URL) and e-mail address of each organization are given first (when available) in each of the entries below, followed by the phone number, postal address, and a brief description of the organization's work or position. When phoning an organization in another country, please locate and use the appropriate country code, which is not included; these codes may vary depending on which country one is calling from. Organizations that do not list e-mail addresses often include forms on their web sites through which e-mail messages may be sent.

GENERAL-PURPOSE ORGANIZATIONS

Center for the Advancement of Energy Markets (CAEM)
URL: http://www.caem.org
Phone: (703) 250-1580
5765-F Burke Center Parkway, PMB333
Burke, VA 22015-2233
Nonprofit corporation chronicles and analyzes changes in domestic and global energy markets caused by technological and policy developments. It encourages market (competition) orientation in restructuring of the electric industry and offers educational programs and publications.

Competitive Enterprise Institute (CEI)
URL: http://www.cei.org
E-mail: info@cei.org
Phone: (202) 331-1010
1001 Connecticut Avenue, NW
Suite 1250
Washington, DC 20036
Public policy and research organization supporting free enterprise and limited government. Issues of concern include competition in electricity markets, use of natural resources, air quality, and global warming.

Energy Foundation (EF)
URL: http://www.ef.org/home.cfm
E-mail: energyfund@ef.org
Phone: (415) 561-6700
1012 Torney Avenue #1
San Francisco, CA 94129
Group of major charitable foundations interested in sustainable energy, especially in regard to electric power, buildings, transportation, national policy, and climate. The foundation focuses on the United States but also has a sustainable energy program in China.

Energy Future Coalition
URL: http://www.energyfuturecoalition.org
E-mail: info@energyfuturecoalition.org
Phone: (202) 463-1947
1225 Connecticut Avenue, NW
4th Floor
Washington, DC 20036
Broad-based alliance seeking energy policy options that business, labor, and environmental groups can all support. The organization stresses economic opportunities and environmental and security improvements it believes will result from changing from fossil fuels to other energy sources.

Energy Information Administration (EIA)
URL: http://www.eia.doe.gov
E-mail: infoctr@eia.doe.gov
Phone: 202-586-8800
EI 30, 1000 Independence Avenue, SW
Washington, DC 20585
Statistical agency, part of the U.S. Department of Energy. It provides data, forecasts, and analyses on all aspects of energy sources and use,

including interaction with the economy and environment, in the United States and the world.

Federal Energy Regulatory Commission (FERC)
URL: http://www.ferc.gov
E-mail: customer@ferc.gov
Phone: (866) 208-3372
888 First Street, NE
Washington, DC 20426
Independent government agency that regulates interstate transmission of natural gas, oil, and electricity. It also regulates natural gas and hydropower projects and wholesale sales of natural gas and electricity in interstate commerce. It oversees environmental matters related to natural gas and hydroelectricity projects and major electricity policy initiatives, and it administers accounting and financial reporting regulations for certain energy-related companies.

International Energy Agency (IEA)
URL: http://www.iea.org
E-mail: info@iea.org
Phone: +33 (1) 405 76500
9, Rue de la Fédération
15739 Paris Cedex 15
France
Intergovernmental body devoted to maintaining energy supply security (for instance by establishment of Strategic Petroleum Reserves), economic growth, and environmental sustainability through policy cooperation of member governments.

National Association of State Energy Officials (NASEO)
URL: http://www.naseo.org
E-mail: information@naseo.org
Phone: (703) 299-8800
1414 Prince Street
Suite 200
Alexandria, VA 22314
Organization of officials from state and territory energy offices, affiliated with the National Governors' Association, aims to shape state and federal energy policy. It keeps government officials informed about the importance of energy to the economy and environment and about the needs and concerns of the states and territories regarding energy issues.

National Center for Appropriate Technology (NCAT)
URL: http://www.ncat.org
E-mail: info@ncat.org
Phone: (406) 494-4572
P.O. Box 3838
Butte, MT 59702
Looks for small-scale, local, and sustainable solutions to poverty and other problems worldwide, including sustainable energy.

National Commission on Energy Policy
URL: http://www.energycommission.org
Phone: (202) 637-0400
1616 H Street, NW
6th Floor
Washington, DC 20006

Bipartisan group of energy experts, supported by large charitable trusts, working to develop a long-term U.S. energy strategy that will promote national security, economic prosperity, and environmental and health safety.

Resources for the Future (RFF)
URL: http://www.rff.org
Phone: (202) 328-5000
1616 P Street, NW
Washington, DC 20036
Independent institute dedicated to analyzing environmental, energy, and natural resource topics, including electricity and climate change.

Sustainable Energy and Economy Network (SEEN)
URL: http://www.seen.org
E-mail: dwysham@seen.org
Phone: (202) 234-9382, Ext. 208
733 15th Street, NW
Suite 1020
Washington, DC 20005
This project of the Institute for Policy Studies (Washington, D.C.) and the Transnational Institute (Amsterdam) works with citizens' groups on global human rights and environmental issues, including energy, climate change, and security concerns caused by competition to obtain fossil fuels. SEEN stresses that energy should be viewed as part of an overall sustainable development strategy. It focuses on investments that international financial institutions and government agencies make in developing countries.

U.S. Department of Energy (DOE)
URL: http://www.energy.gov/engine/content.do?BT_CODE=DOEHOME
E-mail: the.secretary@hq.doe.gov
Phone: (800) 342-5363
1000 Independence Avenue, SW
Washington, DC 20585
Aims to protect national security and the environment by promoting a diverse supply of energy and reliable, affordable, and environmentally sound energy delivery. The department also sponsors research into different types of energy-related science and technology.

U.S. Energy Association (USEA)
URL: http://www.usea.org
Phone: (202) 312-1230
1300 Pennsylvania Avenue
Suite 550
Washington, DC 20004
Association of organizations, corporations, and government agencies that represents interests of the U.S. energy sector to the World Energy Council and elsewhere. It supports increasing domestic energy supplies, including supplies of fossil fuels.

World Business Council for Sustainable Development (WBCSD)
URL: http://www.wbcsd.ch/templates/TemplateWBCSD5/layout.asp?MenuID=1
E-mail: info@wbcsd.org
Phone: (22) 839 3100
4, chemin de Conches

1231 Conches-Geneva
Switzerland
Coalition of international companies committed to sustainable development that includes economic growth, ecological balance, and social progress. The organization believes that business and sustainable development can benefit one another.

World Energy Assessment (WEA)
URL: http://www.undp.org/seed/eap/activities/wea
E-mail: maria.castillo@undp.org
c/o United Nations Development Programme (UNDP)
Sustainable Energy Programme
Environmentally Sustainable Development Group
Bureau for Development Policy
304 East 45th Street
New York, NY 10017
Established by the United Nations Development Programme, the UN Department Economic and Social Affairs, and the World Energy Council to provide information that will help in bringing sustainable energy to the world's people. It evaluates economic, social, security, and environmental issues related to energy and determines the compatibility of different energy sources with objectives in these areas.

World Energy Council (WEC)
URL: http://www.worldenergy.org/wec-geis
E-mail: info@worldenergy.org
Phone: (20) 7734 5996
Regency House
1-4 Warwick Street, 5th Floor
London W1B 5LT
United Kingdom
World's foremost global multi-energy organization, accredited by the United Nations. It is made up of member committees in more than 90 countries, including most of the largest energy producers and consumers. WEC covers all energy sources and is not aligned with governments, commercial interests, or political viewpoints.

ORGANIZATIONS FOCUSING ON CONVENTIONAL FUELS

American Association of Petroleum Geologists (AAPG)
URL: http://www.aapg.org
E-mail: postmaster@aapg.org
Phone: (800) 364-2274
P.O. Box 979
Tulsa, OK 74101-0979
AAPG, the world's largest professional geological society, fosters scientific research and technological development in geology by means of publications, conferences, information distribution, and educational opportunities.

Organizations and Agencies

American Gas Association (AGA)
URL: http://www.aga.org
Phone: (202) 824-7000
400 N. Capitol Street, NW
Suite 450
Washington, DC 20001
Trade association representing most of the local energy utility companies that deliver natural gas to homes, businesses, and industries in the United States.

American Nuclear Society (ANS)
URL: http://www.ans.org
Phone: (708) 352-6611
555 North Kensington Avenue
La Grange Park, IL 60526
Nonprofit, international scientific and educational organization aiming to unify the professional activities of scientists and others in the nuclear industry and to develop and safely apply nuclear science and technology for public benefit.

American Petroleum Institute (API)
URL: http://api-ec.api.org/frontpage.cfm
Phone: (202) 682-8000
1220 L Street, NW
Washington, DC 20005-4070
Primary trade association of the oil and natural gas industry in the United States. Its Energy Consumers web site includes basic information about oil and natural gas, opinions on policy and environmental issues including global warming, educational materials, industry statistics, and more.

Americans for Balanced Energy Choices
URL: http://www.balancedenergy.org/abec
E-mail: membership@balancedenergy.org
Phone: (877) 358-6699
P.O. Box 1638
Alexandria, VA 22313
Organization funded by the coal industry stresses the importance of coal as a fuel for producing electricity.

Association for the Study of Peak Oil and Gas
URL: http://www.peakoil.net
E-mail: aleklett@tsl.uu.se
Phone: (70) 425-0604
Box 25182
SE-750 25 Uppsala
Sweden
Works to determine the date and impact of the peak and decline of world production of oil and natural gas and to warn the world of the serious consequences of this decline.

Canadian Association of Petroleum Producers (CAPP)
URL: http://www.capp.ca
E-mail: communication@capp.ca
Phone: (403) 267-1100
350 7th Avenue, SW
Suite 2100
Calgary, Alberta
Canada T2P 3N9
Trade association representing almost all the companies that explore for, develop, and produce Canada's crude oil, natural gas, and oil sands.

Center for Energy and Economic Development (CEED)
URL: http://www.ceednet.org/ceed
E-mail: info@ceednet.org
Phone: (703) 684-6292
333 John Carlyle Street
Suite 530
Alexandria, VA 22314
Group advocating use of coal as a fuel for generating electricity.

Centre for Global Energy Studies (CGES)
URL: http://www.cges.co.uk
E-mail: marketing@cges.co.uk
Phone: (0)20 7235-4334
17 Knightsbridge
London SW1X 7LY
United Kingdom
Founded in 1990 by Sheikh Ahmed Zaki Yamani, minister for Petroleum and Mineral Resources of Saudi Arabia, to provide independent and objective information and analysis on key energy issues, especially developments related to the oil and natural gas market.

EarthRights International
URL: http://www.earthrights.org
E-mail: infousa@earthrights.org
Phone: (202) 466-5188
1612 K Street, NW
Suite 401
Washington, DC 20006
Uses legal action to defend human rights and the environment against alleged abuses by large corporations, including oil companies, and repressive governments.

Energy Intelligence Group
URL: http://www.energyintel.com
Phone: (212) 532-1112
5 East 37th Street
5th Floor
New York, NY 10016-2807
Gathers data and reports on the international oil and gas business and events that affect it. The group publishes *Oil Daily* and *Petroleum Intelligence Weekly* as well as other data sources and research services.

Independent Petroleum Association of America (IPAA)
URL: http://www.ipaa.org
Phone: (202) 857-4722
1201 15th Street, NW
Suite 300
Washington, DC 20005
National trade association for the exploration and production segment of the U.S. petroleum industry.

Institute of Nuclear Power Operations (INPO)
URL: http://www.eh.doe.gov/inpo
E-mail: esh-infocenter@eh.doe.gov
The nuclear electric utility industry created this organization in 1979 to promote high levels of safety and reliability in operation of nuclear power plants. All companies that operate commercial nuclear plants in the United States are members. INPO is affiliated with the U.S. Department of Energy (DOE).

International Atomic Energy Agency (IAEA)
URL: http://www.iaea.or.at
Phone: (+431) 2600-0
E-mail: Official.Mail@iaea.org
P.O. Box 100
Wagramer Strasse 5
A-1400 Vienna
Austria
Established by the United Nations in 1957 to promote safe, secure, and peaceful nuclear technologies worldwide, this group inspects nuclear material and activities to ensure that they are not used for military purposes, helps countries improve safety and security of nuclear facilities and handle emergencies, and promotes peaceful applications of nuclear science and technology in developing countries.

International Labor Rights Fund
URL: http://www.laborrights.org
E-mail: laborrights@igc.org
Phone: (202) 347-4100
733 15th Street, NW
#920
Washington, DC 20005
Aims to achieve just and humane treatment for workers worldwide. The group has filed suits under the Alien Tort Claims Act against several oil companies for taking part in or being "knowingly complicit" about governmental abuses of human rights in developing countries.

Interstate Natural Gas Association of America (INGAA)
URL: http://www.ingaa.org
Phone: (202) 216-5900
10 G Street, NE
Suite 700
Washington, DC 20002
Trade organization of the natural gas pipeline industry in the United States, Canada, and Mexico.

National Petrochemical and Refiners Association (NPRA)
URL: http://www.npradc.org
E-mail: info@npra.org
Phone: (202) 457-0480
1899 L Street, NW
Suite 1000
Washington, DC 20036-3896
Trade association of U.S. petrochemical and oil refining industries.

National Petroleum Council (NPC)
URL: http://www.npc.org
E-mail: info@npc.org
Phone: (202) 393-6100
1625 K Street, NW
Suite 600
Washington, DC 20006
Policy advisory committee to the U.S. secretary of energy, representing the oil and natural gas industries. It does not otherwise act as a trade association or concern itself with trade practices.

Natural Gas Supply Association (NGSA)
URL: http://www.ngsa.org
Phone: (202) 326-9300
805 15th Street, NW
Suite 510
Washington, DC 20005

Represents producers and marketers of natural gas in the United States. It promotes use of natural gas and competitive markets operating under reasonable regulation.

Nuclear Energy Institute (NEI)
URL: http://www.nei.org
E-mail: webmasterp@nei.org
Phone: (202) 739-8000
1776 I Street NW
Suite 400
Washington, DC 20006-3708
Policy development organization of the nuclear energy and technologies industry, representing the industry's point of view and needs to government and providing information to members, policymakers, the media, and the public.

Nuclear Information and Resource Service (NIRS)
URL: http://www.nirs.org
E-mail: nirsnet@nirs.org
Phone: (202) 328-0002
1424 16th Street, NW, #404
Washington, DC 20036
Group opposed to nuclear power, affiliated with the World Information Service on Energy. The group acts as an information and networking center for individuals and organizations with similar interests.

Nuclear Regulatory Commission (NRC)
Office of Public Affairs
URL: http://www.nrc.gov
E-mail: opa@nrc.gov
Phone: (800) 368-5642
Washington, DC 20555
Government agency that regulates nuclear materials and facilities to protect safety and security, public health, and the environment. NRC formulates policy and regulations for the nuclear industry.

Office of Fossil Energy
URL: http://www.fe.doe.gov
E-mail: fewebmaster@hq.doe.gov
Phone: 202-586-6503
c/o U.S. Department of Energy
Forrestal Building
1000 Independence Avenue, SW
Washington, DC 20585
Part of the U.S. Department of Energy, it works on several presidentially mandated programs to improve fossil fuel technology, including a clean coal program. It also manages the Strategic Petroleum Reserve.

Oil Depletion Analysis Centre (ODAC)
URL: http://www.odac-info.org
E-mail: odac@btconnect.com
Phone: (0)20 7424 0049
140 Fortess Road
London NW5 2HP
United Kingdom
British educational charity working to raise international awareness of coming oil shortages and their effects.

Organization of Petroleum Exporting Countries (OPEC)
URL: http://www.opec.org
Phone: (1) 21112-270
Obere Donaustrasse 93
A-1020 Vienna
Austria

Represents 11 developing countries that rely heavily on oil revenues as their main source of income. OPEC affects the oil market by making collective adjustments to member countries' oil production.

Petroleum Industry Research Foundation, Inc.
URL: http://www.pirinc.org
Nonprofit organization that studies and reports on energy economics with special emphasis on oil. It provides information and expertise to policymakers and the media. The group does not speak for the oil industry or any of its segments.

Petroleum Marketers Association of America (PMAA)
URL: http://www.pmaa.org
E-mail: info@pmaa.org
Phone: 703-351-8000
1901 N. Fort Myer Drive
Suite 500
Arlington, VA 22209-1604
National trade association of petroleum marketers.

Refinery Reform Campaign
URL: http://www.refineryreform.org
E-mail: dannylarson@earthlink.net
Phone: (415) 643-1870
222 Richland Avenue
San Francisco, CA 94110
National campaign to clean up America's oil refineries and reduce dependence on fossil fuels.

World Association of Nuclear Operators (WANO)
URL: http://www.wano.org.uk
Cavendish Court
11-15 Wigmore Street
First Floor
London W1U 1PF
United Kingdom
Established by nuclear plant operators after the Chernobyl accident in 1986 to coordinate efforts to ensure that such a disaster never occurs again.

World Information Service on Energy (WISE)
URL: http://www.antenna.nl/wise
E-mail: wiseamster@antenna.nl
Phone: (20) 612-6368
P.O. Box 59636
1040 LC Amsterdam
The Netherlands
International group opposed to nuclear power. It primarily acts as networker for similar organizations in different countries but also takes part in some direct action.

World Petroleum Congress
URL: http://www.world-petroleum.org
Phone: (0) 20 763-74958
1 Duchess Street
4th Floor
Suite 1
London W1A 3DE
United Kingdom
Provides nonpolitical forum for discussing issues facing the global oil industry. Most oil and gas producing and consuming nations are members.

ORGANIZATIONS FOCUSING ON ALTERNATIVE FUELS AND ENERGY EFFICIENCY

Alliance to Save Energy (ASE)
URL: http://www.ase.org
E-mail: info@ase.org
Phone: (202) 857-0666
1200 18th Street, NW
Suite 900
Washington, DC 20036
Nonprofit coalition of business, government, environmental, and consumer groups that share support for energy efficiency. The alliance conducts research, educational programs, and policy advocacy, designs and implements energy efficiency projects, and promotes development and use of efficient technology.

American Bioenergy Association
URL: http://www.biomass.org
E-mail: genevieve_cullen@biomass.org or shirley_neff@biomass.org
"Voice" of the biomass energy industry in the United States.

American Council for an Energy-Efficient Economy (ACEEE)
URL: http://www.aceee.org
E-mail: info@aceee.org
Phone: (202) 429-8873
1001 Connecticut Avenue, NW
Suite 801
Washington, DC 20036
Nonprofit organization dedicated to advancing energy efficiency as a way of promoting economic prosperity along with environmental protection.

American Solar Energy Society (ASES)
URL: http://www.ases.org
E-mail: ases@ases.org
Phone: (303) 443-3130
2400 Central Avenue
Suite A
Boulder, CO 80301
U.S. section of the International Solar Energy Society, it advances the use of solar energy by organizing a conference and the National Solar Tour, publishing a magazine *(Solar Today)* and newsletter, arranging a Solar Action Network, and more.

American Wind Energy Association (AWEA)
URL: http://www.awea.org
E-mail: windmail@awea.org
Phone: (202) 383-2500
1101 14th Street, NW
12th Floor
Washington, DC 20005
Advocates wind as a reliable, environmentally superior energy source.

Apollo Alliance
URL: http://www.apolloalliance.org
E-mail: feedback@apolloalliance.org
Phone: (202) 955-5665 Ext. 140

c/o Institute for America's Future
1025 Connecticut Avenue
Suite 205
Washington, DC 20036
Stresses that increasing energy efficiency and use of renewable energy sources will create many new jobs as well as protecting the environment.

Canadian Hydropower Association
URL: http://www.canhydropower.org
E-mail: info@canhydropower.org
Phone: (613) 751-6655
255 Albert Street
Suite 600
Ottawa, Ontario K1P 6A9
Canada
Represents the hydropower industry in Canada.

Center for Resource Solutions
URL: http://www.resource-solutions.org
E-mail: mlehman@resource-solutions.org
Phone: (415) 561-2100
Presidio Building 97
P.O. Box 29512
San Francisco, CA 94129
Promotes renewable energy sources and economic and environmental sustainability in the United States and worldwide.

European Renewable Energy Council (EREC)
URL: http://www.erec-renewables.org/default.htm
E-mail: erec@erec-renewables.org
Phone: (2) 546-1933
26, rue du Trône
B-1000 Brussels
Belgium
Umbrella organization for leading European renewable energy industry and research associations, including the European Biomass Industry Association, the European Photovoltaic Industry Association, the European Small Hydropower Association, the European Solar Thermal Industry Federation, and the European Wind Energy Association.

Geothermal Energy Association
URL: http://www.geo-energy.org
E-mail: research@geo-energy.org
Phone: (202) 454-5261
209 Pennsylvania Avenue, SE
Washington, DC 20003
Trade association for the U.S. geothermal industry and those who support this energy source for heating and electricity generation.

Geothermal Resources Council
URL: http://www.geothermal.org
E-mail: grclib@sbcglobal.net
Phone: (530) 758-2360
P.O. Box 1350
2001 Second Street
Suite 5
Davis, CA 95617-1350
Serves as a focal point for continuing professional development of its members and encourages development of geothermal resources and technology worldwide.

International Geothermal Association (IGA)
URL: http://iga.igg.cnr.it/index.php
E-mail: igas@samorka.is
Phone: (354) 588-4437
c/o Samorka
Sudurlandsbraut 48
108 Reykjavik
Iceland
Scientific, educational, and cultural organization intended to encourage research, development, and utilization of geothermal resources. The organization publishes scientific and technical information on geothermal energy and consults with the United Nations and the European Union.

International Hydropower Association (IHA)
URL: http://www.hydropower.org
E-mail: iha@hydropower.org
Phone: (20) 8288-1918
Westmead House
Suite 55
123 Westmead Road
Sutton, Surrey SM1 4JH
United Kingdom
Organization of organizations and individuals working with or studying hydroelectric power, founded in 1995 by the United Nations Educational, Scientific, and Cultural Organization (UNESCO) to advance knowledge of hydropower and promote good industry practice.

International Solar Energy Society (ISES)
URL: http://www.ises.org/ises.nsf!Open
E-mail: hq@ises.org
Phone: (761) 45906-0
Villa Tannheim
Wiesentalstrasse 50
79115 Freiburg
Germany
Encourages the use of renewable energy sources in general and solar energy in particular.

Low Impact Hydropower Institute
URL: http://www.lowimpacthydro.org
E-mail: info@lowimpacthydro.org
Phone: (207) 773-8190
34 Providence Street
Portland, ME 04103
Evaluates hydropower installations in terms of their effects on river flow, water quality, fish passage and protection, threatened and endangered species protection, watershed protection, cultural resource protection, recreation, and facilities recommended for removal. The group certifies facilities that meet its goals in all these areas as "low impact" and environmentally responsible.

National Association of Energy Service Companies
URL: http://www.naesco.org
Phone: (202) 822-0950
1615 M Street, NW

Suite 800
Washington, DC 20036
Trade association for those who provide "energy services" that increase efficiency and reduce demand. The group works to open new markets for such services and persuade policymakers and consumers that use of cost-effective efficiency technologies has substantial economic benefits.

National Hydropower Association
URL: http://www.hydro.org
E-mail: help@hydro.org
Phone: (202) 682-1700
1 Massachusetts Avenue, NW
Suite 850
Washington, DC 20001
Only national trade association dedicated exclusively to representing the interests of the hydropower industry.

Office of Energy Efficiency and Renewable Energy (EERE) and National Renewable Energy Laboratory (NREL)
URLs: http://www.eere.energy.gov/ and http://www.nrel.gov/
Phone: (303) 275-4826
Mail Stop 1521
1617 Cole Boulevard
Golden, CO 80401-3393
Parts of the U.S. Department of Energy. The laboratory does research and development on alternative energy technologies. The office works to spread use of renewable energy sources by forming partnerships with industries and universities to develop and manage projects that transfer renewable technologies to the marketplace.

Rocky Mountain Institute (RMI)
URL: http://www.rmi.org
Phone: (970) 927-3851
1739 Snowmass Creek Road
Snowmass, CO 81654-9199
Founded in 1982 by Amory B. Lovins and L. Hunter Lovins to develop energy-efficient technologies and promote the environmental and economic advantages of increasing energy efficiency. The institute encourages market-oriented rather than governmental solutions to energy and environmental problems.

Solar Energy Industries Association (SEIA)
URL: http://www.seia.org
E-mail: info@seia.org
Phone: (202) 682-0556
805 15th Street
Suite #510
Washington, DC 20005
National trade association of manufacturers and others involved in the solar energy industry.

Solar Energy Research and Education Foundation (SEREF)
URL: http://www.serefonline.org/index2.html
E-mail: info@serefonline.org
4733 Bethesda Avenue
Suite 608
Bethesda, MD 20814

Develops and distributes educational material about solar and other alternative energy sources. The foundation works closely with national renewable energy trade organizations.

Vote Solar Initiative
URL: http://www.votesolar.org
E-mail: david@votesolar
Phone: (415) 874-7435
182 2nd Street
Suite 400
San Francisco, CA 94105
Helps city governments implement large-scale, cost-effective solar projects.

World Wind Energy Association
URL: http://www.wwindea.org
E-mail: secretariat@wwindea.org
Phone: (228) 369-4080
Charles-de-Gaulle-Strasse 5
53113 Bonn
Germany
Promotes and provides information about wind power worldwide.

ORGANIZATIONS FOCUSING ON ELECTRICITY

American Public Power Association (APPA)
URL: http://www.appanet.org
E-mail: mrufe@appanet.org
Phone: (202) 467-2900
2301 M Street, NW
Washington, DC 20037-1484
Service organization for the country's community-owned, nonprofit electric utilities.

Edison Electric Institute (EEI)
URL: http://www.eei.org
E-mail: feedback@eei.org
Phone: (202) 508-5000
701 Pennsylvania Avenue, NW
Washington, DC 20004-2696
Trade association of investor-owned electric utilities that puts out a magazine, *Electric Perspectives,* and position papers on issues including energy infrastructure, environment, energy policy, and transmission reliability.

Electric Power Research Institute (EPRI)
URL: http://www.epri.com
E-mail: askepri@epri.com
3412 Hillview Avenue
Palo Alto, CA 94304
Nonprofit consortium that carries out research and technology development for the benefit of electric utilities and their customers worldwide.

Electric Power Supply Association (EPSA)
URL: http://www.epsa.org/forms/documents/DocumentFormPublic/

E-mail: epsainfo@epsa.org
Phone: (202) 628-8200
1401 New York Avenue, NW
11th Floor
Washington, DC 20005-2110
Trade association for independent (nonutility) electricity generators, power marketers, and other competitive electricity suppliers. The association advocates policies to produce a fully competitive electric power supply marketplace.

North American Electric Reliability Council (NERC)
URL: http://www.nerc.com
E-mail: info@nerc.com
Phone: (609) 452-8060
116-390 Village Boulevard
Princeton, NJ 08540-5731
Voluntary industry organization whose mission is to ensure that the North American electric transmission system is adequate, reliable, and secure.

ORGANIZATIONS FOCUSING ON ENVIRONMENTAL ISSUES

Bluewater Network
URL: http://www.bluewaternetwork.org
E-mail: bluewater@bluewaternetwork.org
Phone: (415) 544-0790
311 California Street
Suite 510
San Francisco, CA 94104
Works to remove root causes of air and water pollution, global warming, and habitat destruction, especially dependence on fossil fuels.

Climate Action Network (United States)
URL: http://www.climatenetwork.org
E-mail: info@climatenetwork.org
Phone: (202) 513-6240
1200 New York Avenue, NW
Suite 1400
Washington, DC 20005
U.S. branch of a worldwide network of nongovernmental organizations working to limit human-induced climate change and protect the atmosphere while allowing sustainable and equitable development.

Electric Reliability Coordinating Council
URL: http://www.electricreliability.org
E-mail: info@electricreliability.org
Coalition of electric utilities and others who feel that the Environmental Protection Agency's requirement that remodeled power plants meet New Source Review standards regarding emission of air pollutants threatens the

reliability of the country's electrical system and unnecessarily increases the cost of power.

Energy Saving Trust (EST)
URL: http://www.est.org.uk
E-mail: media@est.co.uk
Phone: (020) 7222-0101
21 Dartmouth Street
London SW1H 9BP
United Kingdom
Established by the British government, the Energy Saving Trust is one of the country's leading organizations addressing the damaging effects of climate change. It devises ways for households, small businesses, and vehicle drivers to use energy more efficiently and reduce emissions of carbon dioxide and other pollutants.

Environmental and Energy Study Institute (EESI)
URL: http://www.eesi.org/index.html
E-mail: eesi@eesi.org
Phone: (202) 628-1400
122 C Street, NW
Suite 630
Washington, DC 20001-2109
Provides information and policy initiatives aimed at promoting environmentally sustainable societies. Energy-related research topics include renewable energy sources, energy efficiency, global climate changes, biofuels, and clean bus technologies.

Global Environment Facility (GEF)
URL: http://www.gefweb.org
E-mail: secretariat@TheGEF.org
Phone: (202) 473-0508
1818 H Street, NW
Washington, DC 20433
Independent financial organization that provides grants to developing countries for projects that benefit the environment and promote sustainable livelihoods, including projects that involve renewable energy sources and reduction of carbon dioxide emissions.

Greening Earth Society
URL: http://www.greeningearthsociety.org
E-mail: info@co2andclimate.org
Phone: (800) 529-4503
333 John Carlyle Street
Suite 530
Alexandria, VA 22314
Association of rural electric cooperatives, municipal electric utilities, and others who are skeptical about claims that the planet's climate is endangered by humanity's emissions of carbon dioxide.

Institute for Energy and Environmental Research (IEER)
URL: http://www.ieer.org
E-mail: ieer@ieer.org
Phone: (301) 270-5500
6935 Laurel Avenue
Suite 201
Takoma Park, MD 20912
Aims to provide understandable and accurate scientific and technical information on energy and environmental issues.

Intergovernmental Panel on Climate Change (IPCC)
URL: http://www.ipcc.ch
E-mail: IPCC-Sec@wmo.int
Phone: (22) 730-8208
c/o World Meteorological Organization
7 bis Avenue de la Paix
C.P. 2300
CH-1211
Geneva 2
Switzerland
Established by the World Meteorological Organization and the United Nations Environmental Program (UNEP) to assess scientific, technical, and socioeconomic information relevant to the understanding of climate change, its potential impacts, and options for adaptation and mitigation.

Pew Center on Global Climate Change
URL: http://www.pewclimate.org
Phone: (703) 516-4146
2101 Wilson Boulevard
Suite 550
Arlington, VA 22201
Brings together scientists, policymakers, business leaders, and other experts to analyze this complex issue and seek ways to protect the climate while sustaining economic growth. It produces numerous reports. Pew works with major corporations who are members of its Business Environmental Leadership Council to implement market-based mechanisms to reduce greenhouse gas emissions.

Republicans for Environmental Protection
URL: http://www.repamerica.org
E-mail: info@repamerica.org
Phone: (505) 889-4544
3200 Carlisle Boulevard
Suite 228
Albuquerque, NM 87110
Republicans who promote a healthy environment combined with a sound economy. Their concerns include energy issues. They favor some increased drilling for fossil fuels on public lands but also conservation, energy efficiency, and increased use of renewable energy sources.

Royal Commission on Environmental Pollution (RCEP)
URL: http://www.rcep.org.uk
E-mail: enquiries@rcep.org.uk
Phone: (0)20 7799 8970
The Sanctuary
Third Floor
Westminster, London SW1P 3JS
United Kingdom
Independent advisory body established by the British government, it produces reports such as "Biomass as a Renewable Energy Source," published in May 2004.

Union of Concerned Scientists
URL: http://www.ucsusa.org
Phone: (617) 547-5552
2 Brattle Square
Cambridge, MA 02238-9105
Scientist-environmentalist group whose areas of concern include clean energy, clean vehicles, and

climate change and other threats to the global environment.

U.S. Environmental Protection Agency (EPA)
URL: http://www.epa.gov
Phone: (202) 272-0167
Ariel Rios Building
1200 Pennsylvania Avenue, NW
Washington, DC 20460
Federal government agency with a mission to protect the environment and human health. It administers the Clean Air Act and has jurisdiction over some other areas in which environmental and energy regulation intersect.

U.S. Global Change Research Information Office (GCRIO)
URL: http://www.gcrio.org
E-mail: information@gcrio.org
Phone: (202) 223-6262
1717 Pennsylvania Avenue, NW
Suite 250
Washington, DC 20006
Government agency that disseminates scientific information useful in preventing, minimizing, or adapting to the effects of global climate change.

World Resources Institute (WRI)
URL: http://www.wri.org
E-mail: swilson@wri.org
Phone: (202) 729-7600
10 G Street, NE
Suite 800
Washington, DC 20002
Explores issues at the intersection of environmental protection and economic development, including energy and climate change.

Worldwatch Institute
URL: http://www.worldwatch.org
E-mail: worldwatch@worldwatch.org
Phone: (202) 452-1999
1776 Massachusetts Avenue, NW
Washington, DC 20036-1904
Provides information on interactions among key environmental, economic, and social trends, with the goal of bringing about a transition to an environmentally sustainable and socially just society worldwide. Energy is a major research area.

PART III

APPENDICES

APPENDIX A

TABLES, GRAPHS, AND MAPS

PAST AND PROJECTED FUTURE WORLD ENERGY CONSUMPTION

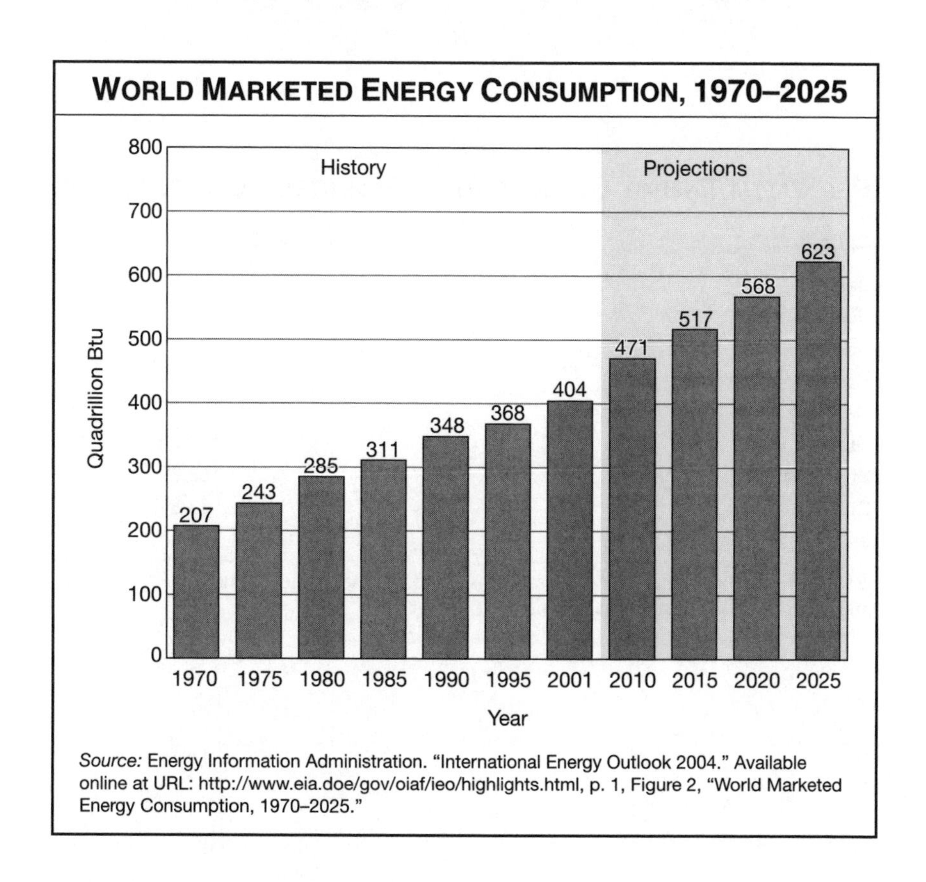

Source: Energy Information Administration. "International Energy Outlook 2004." Available online at URL: http://www.eia.doe/gov/oiaf/ieo/highlights.html, p. 1, Figure 2, "World Marketed Energy Consumption, 1970–2025."

World's Largest Total Primary Energy Consumers, 2002

Country	Quadrillion British Thermal Units
United States	97.649
China	43.177
Russia	27.536
Japan	21.965
Germany	14.269
India	13.981
Canada	13.065
France	10.986
United Kingdom	9.581
Brazil	8.591

Source: Energy Information Administration

Per Capita Energy Consumption in Selected Countries, 2002

Country	Million British Thermal Units
Canada	417.8
United States	339.1
Russia	191.1
France	183.6
Germany	173.1
Japan	172.3
United Kingdom	162.2
China	33.3
India	13.3

Source: Energy Information Administration, World Per Capita Total Primary Energy Consumption, 1980–2002. Posted in June 7, 2004.

Appendix A

PAST AND PREDICTED WORLD ENERGY CONSUMPTION BY SOURCE

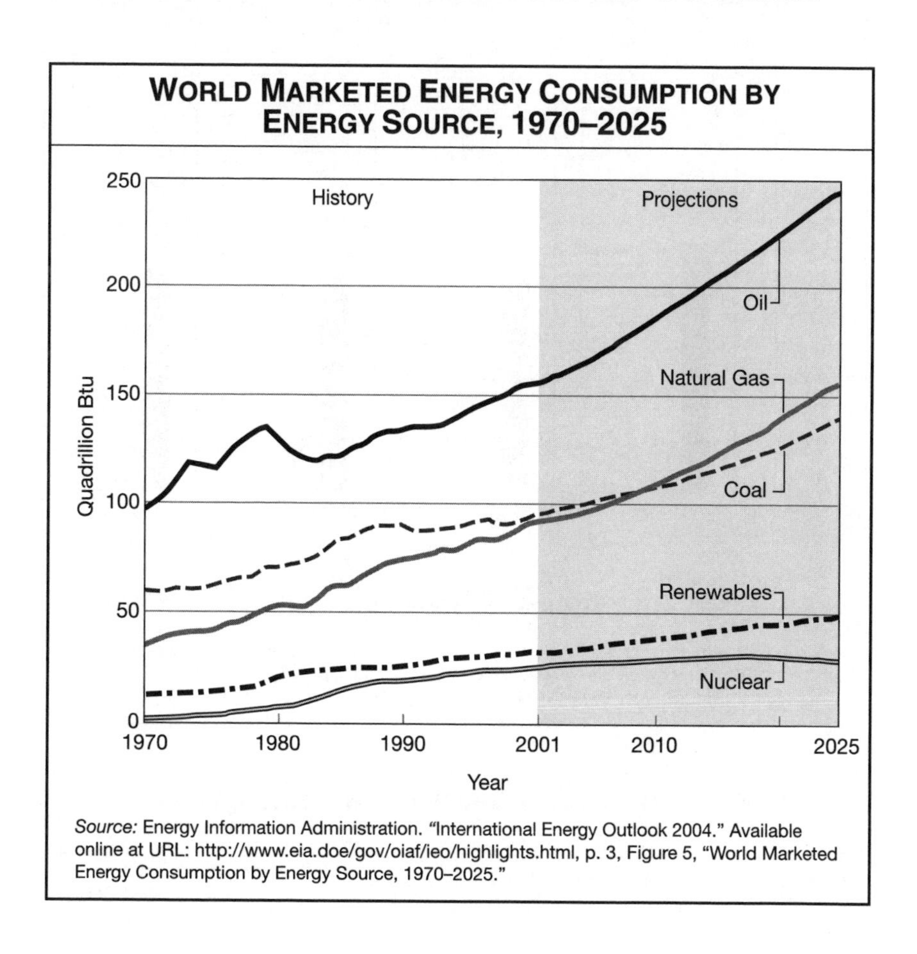

Source: Energy Information Administration. "International Energy Outlook 2004." Available online at URL: http://www.eia.doe/gov/oiaf/ieo/highlights.html, p. 3, Figure 5, "World Marketed Energy Consumption by Energy Source, 1970–2025."

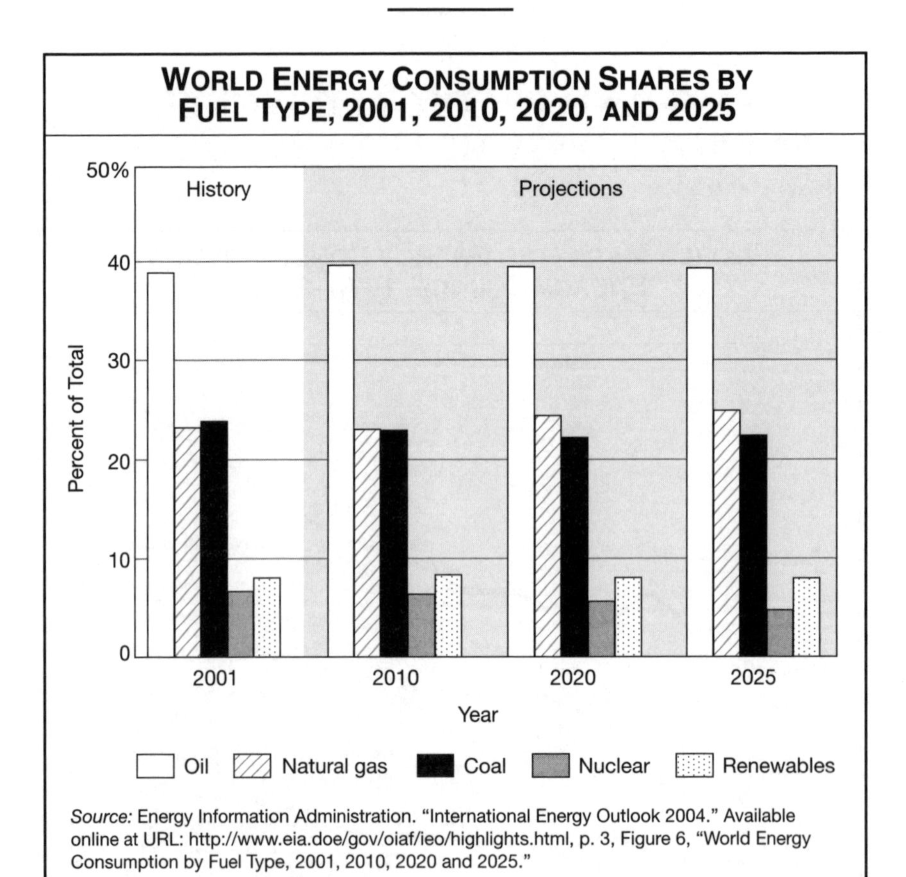

Source: Energy Information Administration. "International Energy Outlook 2004." Available online at URL: http://www.eia.doe/gov/oiaf/ieo/highlights.html, p. 3, Figure 6, "World Energy Consumption by Fuel Type, 2001, 2010, 2020 and 2025."

WORLD'S LARGEST CRUDE OIL RESERVES, 2003

Country	Billion Barrels
Saudi Arabia	261.800
Canada	180.021
Iraq	112.500
United Arab Emirates	97.800
Kuwait	96.500
Iran	89.700
Venezuela	77.800
Russia	60.000
Libya	29.500
United States	22.677

Source: Oil and Gas Journal, reprinted by Energy Information Administration

Appendix A

WORLD CARBON DIOXIDE INTENSITY BY SELECTED COUNTRIES AND REGIONS, 2001 AND 2025

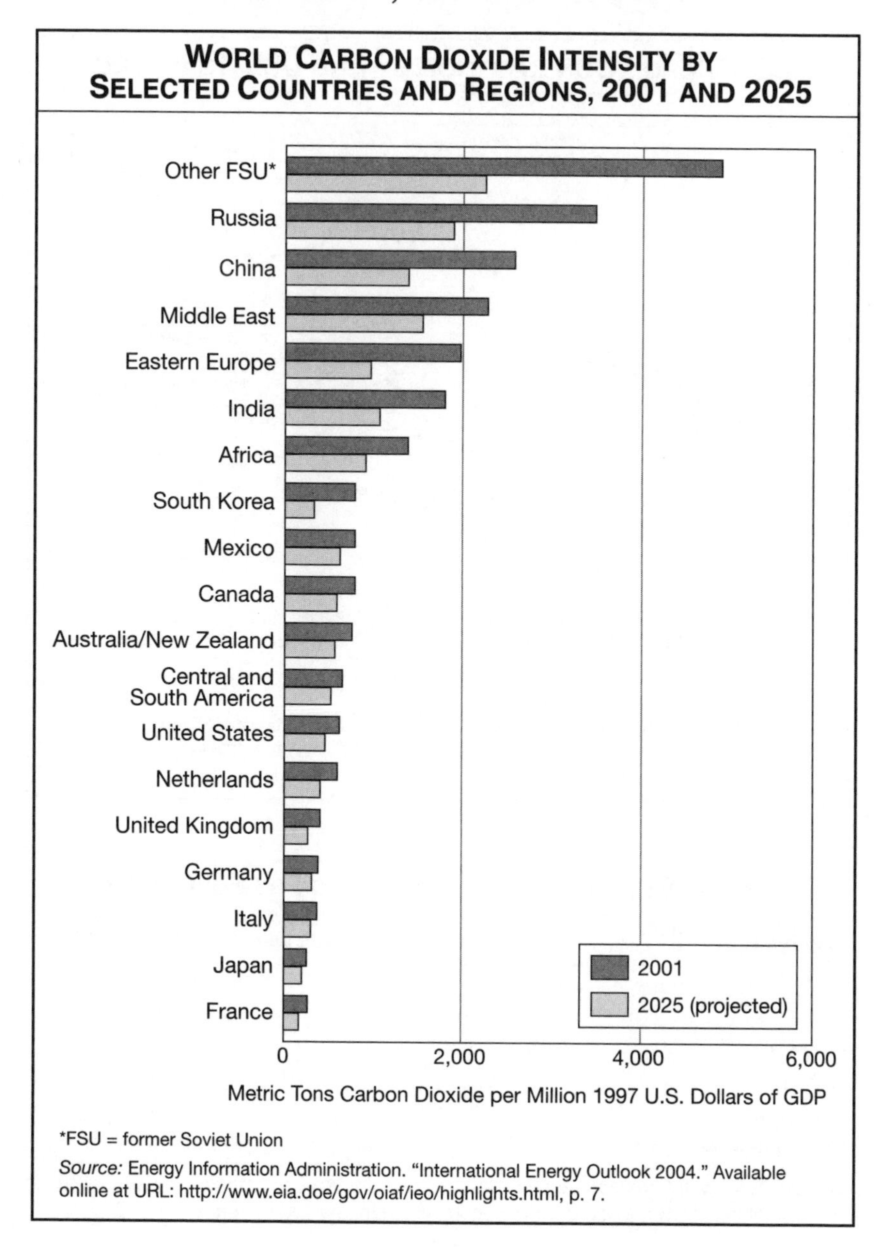

*FSU = former Soviet Union

Source: Energy Information Administration. "International Energy Outlook 2004." Available online at URL: http://www.eia.doe/gov/oiaf/ieo/highlights.html, p. 7.

UNITED STATES CONSUMPTION OF ENERGY BY SECTOR AND YEAR, 1950–2000

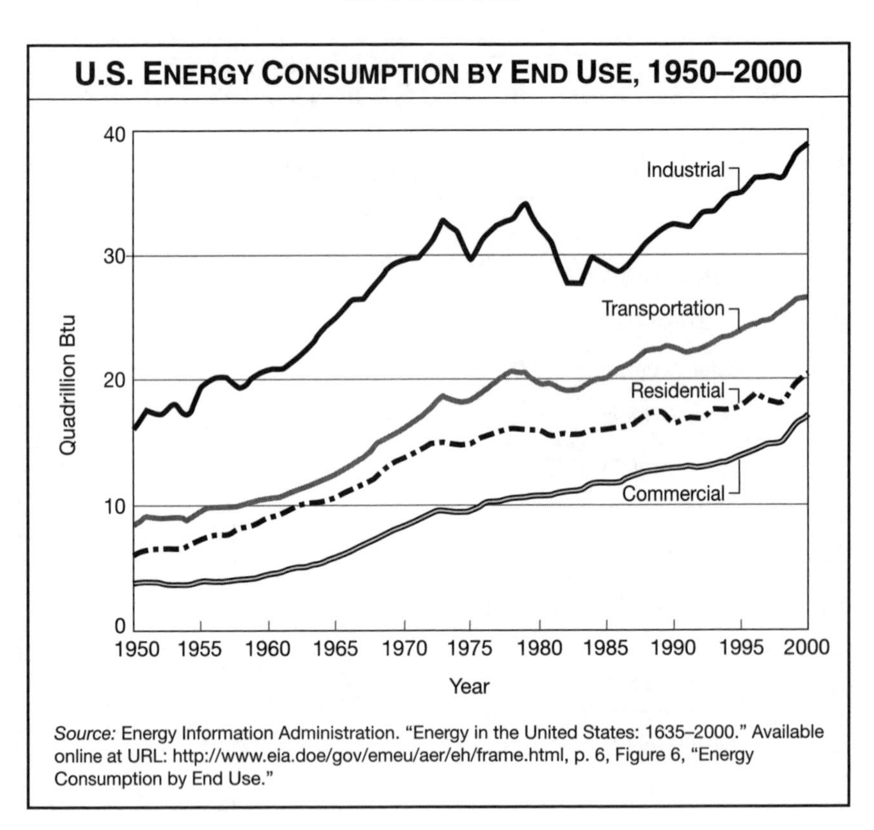

Source: Energy Information Administration. "Energy in the United States: 1635–2000." Available online at URL: http://www.eia.doe/gov/emeu/aer/eh/frame.html, p. 6, Figure 6, "Energy Consumption by End Use."

UNITED STATES ENERGY CONSUMPTION BY FUEL, 2003

Fuel Type	Percentage of Overall Consumption
Oil	40 percent
Coal	23 percent
Natural gas	23 percent
Nuclear	8 percent
Hydropower	3 percent
Other renewables	3 percent

Source: Energy Information Administration, "Country Analysis Briefs: United States of America," http://www.eia.doe.gov/emeu/cabs/usa.html, posted in April 2004, p. 21.

Appendix A

United States Daily Average Oil and Oil Product Imports by Country of Origin, 2004

Country	Thousand Barrels per Day
Canada	1,982
Mexico	1,647
Venezuela	1,443
Saudi Arabia	1,865
Nigeria	1,236
Iraq	816
Algeria	536
United Kingdom	274
Angola	354
U.S. Virgin Islands	355

Source: Energy Information Administration, "Imports of Crude Oil and Petroleum Products into the United States by Country of Origin," last updated August 2004.

Cost of Electricity in the United States by Generating Fuel, 2001

Energy Resource	Cent per Kilowatt Hour
Solar thermal	1
Solar photovoltaic	2
Geothermal	2–8
Natural gas	3.9–4.4
Wind	4–6
Coal	4.8–5.5
Hydroelectric	5.1–11.3
Biomass	5.8–11.6
Nuclear	11.1–14.5

Source: Paula Berinstein, *Alternative Energy: Facts, Statistics, and Issues* (Westport, Conn.: Oryx Press, 2001)

APPENDIX B

STANDARD OIL COMPANY OF NEW JERSEY V. UNITED STATES, 221 U.S. 1, 1911

[*Note*: This court case has been excerpted here. Footnotes and most citations, as well as other material, have been omitted.]

Argued March 14, 15, and 16, 1910.
Ordered for reargument April 11, 1910.
Reargued January 12, 13, 16, . . . , 1911.

Mr. Chief Justice White delivered the opinion of the court:

The Standard Oil Company of New Jersey and thirty-three other corporations, John D. Rockefeller, William Rockefeller, and five other individual defendants, prosecute this appeal to reverse a decree of the court below. Such decree was entered upon a bill filed by the United States under authority of 4 of the act of July 2, 1890, known as the [Sherman] antitrust act, and had for its object the enforcement of the provisions of that act. The record is inordinately voluminous, consisting of twenty-three volumes of printed matter, aggregating about 12,000 pages, containing a vast amount of confusing and conflicting testimony relating to innumerable, complex, and varied business transactions, extending over a period of nearly forty years. In an effort to pave the way to reach the subjects which we are called upon to consider, we propose at the outset, following the order of the bill, to give the merest possible outline of its contents, to summarize the answer, to indicate the course of the trial, and point out briefly the decision below rendered.

The bill and exhibits, covering 170 pages of the printed record, was filed on November 15, 1906. Corporations known as Standard Oil Company of New Jersey [and others] . . . were named as defendants. The bill . . . sought relief upon the theory that the various defendants were engaged in conspiring "to restrain the trade and commerce in petroleum, commonly called 'crude oil,' in refined oil, and in the other products of petroleum, among the several states and territories of the United States and the District of Columbia and with foreign nations, and to monopolize the said commerce." The conspiracy was alleged to have been formed in or about the year 1870 by three of the individual defendants, viz.: John D. Rockefeller, William Rockefeller, and Henry M. Flagler. The detailed averments concerning the alleged conspiracy were arranged with reference to three periods, the first from 1870 to 1882, the second from 1882 to 1899, and the third from 1899 to the time of the filing of the bill.

The general charge concerning the period from 1870 to 1882 was as follows: "That during said first period the said individual defendants, in connection with the Standard Oil Company of Ohio, purchased and obtained interests through stock ownership and otherwise in, and entered into agreements with, various persons, firms, corporations, and limited partnerships engaged in purchasing, shipping, refining, and selling petroleum and its products among the various states, for the purpose of fixing the price of crude and refined oil and the products thereof, limiting the production thereof, and controlling the transportation therein, and thereby restraining trade and commerce among the several states, and monopolizing the said commerce."

To establish this charge it was averred that John D. and William Rockefeller and several other named individuals, who, prior to 1870, composed three separate partnerships engaged in the business of refining crude oil and shipping its products in interstate commerce, organized in the year 1870 a corporation known as the Standard Oil Company of Ohio, and transferred to that company the business of the said partnerships, the members thereof becoming, in proportion to their prior ownership, stockholders in the corporation. It was averred that the other individual defendants soon afterwards became participants in the illegal combination, and either transferred property to the corporation or to individuals, to be held for the benefit of all parties in interest in proportion to their respective interests in the combination; that is, in proportion to their stock ownership in the Standard Oil Company of Ohio. By the means thus stated, it was charged that by the year 1872, the combination had acquired substantially all but three or four of the thirty-five or forty oil refineries located in Cleveland, Ohio. By reason of the power thus obtained, and in further execution of the intent and purpose to restrain trade and to monopolize the commerce, interstate as well as intrastate, in petroleum and its products, the bill alleged that the combination and its members obtained large preferential rates and rebates in many and devious ways over their competitors from various railroad companies, and that by means of the advantage thus obtained many, if not virtually all, competitors were forced either to become members of the combination or were driven out of business; and thus, it was alleged, during the period in question, the following results were brought about: (a) That the combination, in addition to the refineries in Cleveland which it had acquired, as previously stated, and which it had either dismantled to limit production, or continued to operate, also from time to time acquired a large number of refineries of crude petroleum, situated in New York, Pennsylvania, Ohio, and elsewhere. The properties thus acquired, like those previously obtained, although belonging to and being held for the benefit of the combination, were ostensibly divergently controlled, some of them being put in the name of the Standard Oil Company of Ohio, some in the name of corporations or limited partnerships affiliated therewith, or some being left in the name of the original owners, who had become stockholders in the Standard Oil Company of Ohio, and thus members of the alleged illegal combination. (b) That the combination had obtained control of the pipe lines available for transporting oil from the oil fields to the refineries in Cleveland, Pittsburg, Titusville, Philadelphia, New York, and New Jersey. (c) That the combination during the period named had obtained a complete mastery over the Oil industry, controlling 90 per cent of the business of producing, shipping, refining, and selling petroleum and its products, and thus was able to fix the price of crude and refined petroleum, and to restrain and monopolize all interstate commerce in those products.

The averments bearing upon the second period (1882 to 1899) had relation to the claim:

> *That during the said second period of conspiracy the defendants entered into a contract and trust agreement, by which various independent firms, corporations, limited partnerships, and individuals engaged in purchasing, transporting, refining, shipping, and selling oil and the products thereof among the various states, turned over the management of their said business, corporations, and limited partnerships to nine trustees, composed chiefly of certain individuals defendant herein, which said trust agreement was in restraint of trade and commerce, and in violation of law, as hereinafter more particularly alleged.*

The trust agreement thus referred to was set out in the bill. It was made in January, 1882. By its terms the stock of forty corporations, including the Standard Oil Company of Ohio, and a large quantity of various properties which had been previously acquired by the alleged combination, and which was held in diverse forms, as we have previously indicated, for the benefit of the members of the combination, was vested in the trustees and their successors, "to be held for all parties in interest jointly." In the body of the trust agreement was contained a list of the various individuals and corporations and limited partnerships whose stockholders and members, or a portion thereof, became parties to the agreement. . . . The agreement made provision for the method of controlling and managing the property by the trustees, for the formation of additional manufacturing, etc., corporations in various states, and the trust, unless terminated by a mode specified, was to continue "during the lives of the survivors and survivor of the trustees named in the agreement and for twenty-one years thereafter." The agreement provided for the issue of Standard Oil Trust certificates to represent the interest arising under the trust in the properties affected by the trust, which, of course, in view of the provisions of the agreement and the subject to which it related caused the interest in the certificates to be coincident with and the exact representative of the interest in the combination, that is, in the Standard Oil Company of Ohio. Soon afterwards it was alleged the trustees organized the Standard Oil Company of New Jersey and the Standard Oil Company of New York. . . . The bill alleged "that pursuant to said trust agreement the said trustees caused to be transferred to themselves the stocks of all corporations and limited partnerships named in said trust agreement, and caused various of the individuals and copartnerships who owned apparently independent refineries and other properties employed in the business of refining and transporting and selling oil in and among said various states and territories of the United States, as aforesaid, to transfer their property situated in said several states to the respective Standard Oil Companies of said states of New York, New Jersey, Pennsylvania, and Ohio, and other corporations organized or acquired by said trustees from time to time. . . ." For the stocks and property so acquired the trustees issued trust certificates. It was alleged that in 1888 the trustees "unlawfully controlled the stock and ownership of various corporations and limited partnerships engaged in such purchase and transportation, refining, selling, and shipping of oil." . . . The bill charged that during the second period quo warranto proceedings were commenced against the Standard Oil Company of Ohio, which resulted in the entry by the supreme court of Ohio, on March 2, 1892, of a decree. . . .

It was alleged that shortly after this decision, seemingly for the purpose of complying therewith, voluntary proceedings were had apparently to dissolve the trust, but that these proceedings were a subterfuge and a sham because they simply amounted to a transfer of the stock held by the trust in sixty-four of the companies which it controlled to some of the remaining twenty companies, it having controlled before the decree eighty-four in all, thereby, while seemingly in part giving up its dominion, yet in reality preserving the same by means of the control of the companies as to which it had retained complete authority. It was charged that especially was this the case, as the stock in the companies selected for transfer was virtually owned by the nine trustees or the members of their immediate families or associates. The bill further alleged that in 1897 the attorney general of Ohio instituted contempt proceedings in the quo warranto case, based upon the claim that the trust had not been dissolved as required by the decree in that case. . . .

The result of these proceedings, the bill charged, caused a resort to the alleged wrongful acts asserted to have been committed during the third period, as follows:

> *That during the third period of said conspiracy, and in pursuance thereof, the said individual defendants operated through the Standard Oil Company of New Jersey, as a holding corporation, which corporation obtained and acquired the majority of the stocks of the various corporations engaged in purchasing, transporting, refining, shipping, and selling oil into and among the various states and territories of the United States and the District of Columbia and with foreign nations, and thereby managed and controlled the same, in violation of the laws of the United States, as hereinafter more particularly alleged. . . .*

[I]t was alleged in the bill that shortly after these proceedings the trust came to an end, the stock of the various corporations which had been controlled by it being transferred by its holders to the Standard Oil Company of New Jersey. . . .

Reiterating in substance the averments that both the Standard Oil Trust from 1882 to 1899, and the Standard Oil Company of New Jersey, since 1899, had monopolized and restrained interstate commerce in petroleum and its products, the bill at great length additionally set forth various means by which, during the second and third periods, in addition to the effect occasioned by the combination of alleged previously independent concerns, the monopoly and restraint complained of were continued. Without attempting to follow the elaborate averments on these subjects, spread over fifty-seven pages of the printed record, it suffices to say that such averments may properly be grouped under the following heads: Rebates, preferences, and other discriminatory practices in favor of the combination by railroad companies; restraint and monopolization by control of pipe lines, and unfair practices against competing pipe lines; contracts with competitors in restraint of trade; unfair methods of competition, such as local price cutting at the points where necessary to suppress competition; espionage of the business of competitors, the operation of bogus independent companies, and payment of rebates on oil, with the like intent; the division of the United States into districts, and the limiting the operations of the various subsidiary corporations as to such districts so that competition in the sale of petroleum products between such corporations had been entirely eliminated and destroyed; and finally reference was made to what was alleged to be the "enormous and unreasonable profits" earned by the Standard Oil Trust and the Standard Oil Company as a result of the alleged monopoly; which presumably was averred as a means of reflexly inferring the scope and power acquired by the alleged combination.

Coming to the prayer of the bill, it suffices to say that in general terms the substantial relief asked was, first, that the combination in restraint of interstate trade and commerce, and which had monopolized the same, as alleged in the bill, be found to have existence, and that the parties thereto be perpetually enjoined from doing any further act to give effect to it; second, that the transfer of the stocks of the various corporations to the Standard Oil Company of New Jersey, as alleged in the bill, be held to be in violation of the 1st and 2d sections of the anti-trust act, and that the Standard Oil Company of New Jersey be enjoined and restrained from in any manner continuing to exert control over the subsidiary corporations by means of ownership of said stock or otherwise; third, that specific relief by injunction be awarded against further violation of the statute by any of the acts specifically complained of in the bill. There was also a prayer for general relief. . . .

Certain of the defendants filed separate answers, and a joint answer was filed on behalf of the Standard Oil Company of New Jersey and numerous of the other defendants. The scope of the answers will be adequately indicated by quoting a summary on the subject, made in the brief for the appellants.

It is sufficient to say that, whilst admitting many of the alleged acquisitions of property, the formation of the so-called trust of 1882, its dissolution in 1892, and the acquisition by the Standard Oil Company of New Jersey of the stocks of the various corporations in 1899, they deny all the allegations respecting combinations or conspiracies to restrain or monopolize the oil trade; and particularly that the socalled trust of 1882, or the acquisition of the shares of the defendant companies by the Standard Oil Company of New Jersey in 1899, was a combination of independent or competing concerns or corporations. The averments of the petition respecting the means adopted to monopolize the oil trade are traversed either by a denial of the acts alleged, or of their purpose, intent, or effect.

On June 24, 1907, . . . a special examiner was appointed to take the evidence, and his report was filed March 22, 1909. It was heard on April 5 to 10, 1909, . . . before a circuit court consisting of four judges.

The court decided in favor of the United States. In the opinion delivered, all the multitude of acts of wrongdoing charged in the bill were put aside, in so far as they were alleged to have been committed prior to the passage of the anti-trust act, "except as evidence of their (the defendants') purpose, of their continuing conduct, and of its effect."

By the decree which was entered it was adjudged that the combining of the stocks of various companies in the hands of the Standard Oil Company of New Jersey in 1899 constituted a combination in restraint of trade and also an attempt to monopolize and a monopolization under 2 of the anti-trust act. . . .

The Standard Oil Company of New Jersey was enjoined from voting the stocks or exerting any control over the said thirty-seven subsidiary companies, and the subsidiary companies were enjoined from paying any dividends as to the Standard Company, or permitting it to exercise any control over them by virtue of the stock ownership or power acquired by means of the combination. The individuals and corporations were also enjoined from entering into or carrying into effect any like combination which would evade the decree. Further, the individual defendants, the Standard Company, and the thirty-seven subsidiary corporations, were enjoined from engaging or continuing in interstate commerce in petroleum or its products during the continuance of the illegal combination. . . .

We are thus brought face to face with the merits of the controversy.

Both as to the law and as to the facts, the opposing contentions pressed in the argument are numerous, and in all their aspects are so irreconcilable that it is difficult to reduce them to some fundamental generalization, which, by being disposed of, would decide them all. For instance, as to the law. While both sides agree that the determination of the controversy rests upon the correct construction and application of the 1st and second sections of the anti-trust act, yet the views as to the meaning of the act are as wide apart as the poles, since there is no real point of agreement on any view of the act. . . .

So also is it as to the facts. Thus, on the one hand, with relentless pertinacity and minuteness of analysis, it is insisted that the facts establish that the assailed combination took its birth in a purpose to unlawfully acquire wealth by oppressing the public and destroying the just rights of others, and that its entire career exemplifies an inexorable carrying out of such wrongful intents, since, it is asserted, the pathway of the combination from the beginning to the time of the filing of the bill is marked with constant proofs of wrong inflicted upon the public, and is strewn with the wrecks resulting from crushing out, without regard to law, the individual rights of others. Indeed, so conclusive, it is urged, is the proof on these subjects, that it is asserted that the existence of the principal corporate defendant—the Standard Oil Company of New Jersey—with the vast accumulation of property which it owns or controls, because of its infinite potency for harm and the dangerous example which its continued existence affords, is an open and enduring menace to all freedom of trade, and is a byword and reproach to modern economic methods. On the other hand, in a powerful analysis of the facts, it is insisted that they demonstrate that the origin and development of the vast business which the defendants control was but the result of lawful competitive methods, guided by economic genius of the highest order, sustained by courage, by a keen insight into commercial situations, resulting in the acquisition of great wealth, but at the same time serving to stimulate and increase production, to widely extend the distribution of the products of petroleum at a cost largely below that which would have otherwise prevailed, thus proving to be at one and the same time a benefaction to the general public as well as of enormous advantage to individuals. It is not denied that in the enormous volume of proof contained in the record in the period of almost a lifetime, to which that proof is addressed, there may be found acts of wrongdoing, but the insistence is that they were rather the exception than the rule, and in most cases were either the result of too great individual zeal in the keen rivalries of business, or of the methods and habits of dealing which, even if wrong, were commonly practised at the time. And to discover and state the truth concerning these contentions both arguments call for the analysis and weighing, as we have said at the outset, of a jungle of conflicting testimony covering a period of forty years—a duty difficult to rightly perform, and, even if satisfactorily accomplished, almost impossible to state with any reasonable regard to brevity.

Duly appreciating the situation just stated, it is certain that only one point of concord between the parties is discernible, which is, that the controversy in every aspect is controlled by a correct conception of the meaning of the 1st and 2d sections of the anti-trust act. We shall therefore—departing from what otherwise would be the natural order of analysis—make this one point of harmony the initial basis of our examination of the contentions, relying upon the conception that by doing so some harmonious resonance may result adequate to dominate and control the discord with which the case abounds. That is to say, we shall first come to consider the meaning of the 1st and 2d sections of the anti-trust act by the text, and after discerning what by that process appears to be its true meaning, we shall proceed to consider the respective contentions of the parties concerning the act, the strength or weakness of those contentions, as well as the accuracy of the meaning of the act as deduced from the text in the light of the prior decisions of this court concerning it. When we have done this, we shall then approach the facts. Following this course, we shall make our investigation under four separate headings: First. The text of the 1st and 2d sections of the act, originally considered, and its meaning in the light of the common law and the law of this country at the time of its adoption. Second. The contentions of the parties concerning the act, and the scope and effect of the decisions of this court upon which they rely. Third. The application of the statute to facts; and, Fourth. The remedy, if any, to be afforded as the result of such application.

First. The text of the act and its meaning.
We quote the text of the 1st and 2d sections of the act, as follows:

Section 1. Every contract, combination in the form of trust or otherwise, or conspiracy, in restraint of trade or commerce among the several states or with foreign nations, is hereby declared to be illegal. Every person who shall make any such contract, or engage in any such combination or conspiracy, shall be deemed guilty of a misdemeanor, and, on conviction thereof, shall be punished by fine not exceeding $5,000, or by imprisonment not exceeding one year, or by both said punishments, in the discretion of the court.

Section 2. Every person who shall monopolize, or attempt to monopolize, or combine or conspire with any other person or persons to monopolize, any part of the trade or commerce among the several states, or with foreign nations, shall be deemed guilty of a misdemeanor, and, on conviction thereof, shall be punished by fine not exceeding $5,000, or by imprisonment not exceeding one year, or by both said punishments, in the discretion of the court.

. . . As to the 1st section, the words to be interpreted are: "Every contract, combination in the form of trust or otherwise, or conspiracy in restraint of trade or commerce . . . is hereby declared to be illegal." As there is no room for dispute that the statute was intended to formulate a rule for the regulation of interstate and foreign commerce, the question is, What was the rule which it adopted?

In view of the common law and the law in this country as to restraint of trade, which we have reviewed, and the illuminating effect which that history must have under the rule to which we have referred, we think it results:

a. That the context manifests that the statute was drawn in the light of the existing practical conception of the law of restraint of trade, because it groups as within that class, not only contracts which were in restraint of trade in the subjective sense, but all contracts or acts which theoretically were attempts to monopolize, yet which in practice had come to be considered as in restraint of trade in a broad sense.

b. That in view of the many new forms of contracts and combinations which were being evolved from existing economic conditions, it was deemed essential by an all-embracing enumeration to make sure that no form of contract or combination by which an undue restraint of interstate or foreign commerce was brought about could save such restraint from condemnation. The statute under this view evidenced the intent not to restrain the right to make and enforce contracts, whether resulting from combinations or otherwise, which did not unduly restrain interstate or foreign commerce, but to protect that commerce from being restrained by methods, whether old or new, which would constitute an interference—that is, an undue restraint.

c. And as the contracts or acts embraced in the provision were not expressly defined, since the enumeration addressed itself simply to classes of acts, those classes being broad enough to embrace every conceivable contract or combination which could be made concerning trade or commerce or the subjects of such commerce, and thus caused any act done by any of the enumerated methods anywhere in the whole field of human activity to be illegal if in restraint of trade, it inevitably follows that the provision necessarily called for the exercise of judgment which required that some standard should be resorted to for the purpose of determining whether the prohibition contained in the statute had or had not in any given case been violated. Thus not specifying, but indubitably contemplating and requiring a standard, it follows that it was intended that the standard of reason which had been applied at the common law and in this country in dealing with subjects of the character embraced by the statute was intended to be the measure used for the purpose of determining whether, in a given case, a particular act had or had not brought about the wrong against which the statute provided.

And a consideration of the text of the 2d section serves to establish that it was intended to supplement the 1st, and to make sure that by no possible guise could the public policy embodied in the 1st section be frustrated or evaded. The prohibition of the 2d embrace "every person who shall monopolize, or attempt to monopolize, or combine or conspire with any other person or persons to monopolize, any part of the trade or commerce among the several states or with foreign nations. . . ." [T]he word "person" clearly implies a corporation as well as an individual. . . .

Third. The facts and the application of the statute to them.

Beyond dispute the proofs establish substantially as alleged in the bill the following facts:

1. The creation of the Standard Oil Company of Ohio.

2. The organization of the Standard Oil Trust of 1882, and also a previous one of 1879. . . .

The vast amount of property and the possibilities of far-reaching control which resulted from the facts last stated are shown by the statement which we have previously annexed concerning the parties to the trust agreement of 1882. . . . [T]he New Jersey corporation . . . operated to destroy the "potentiality of competition" which otherwise would have existed to such an extent as to cause the transfers of stock which were made to the New Jersey Corporation and the control which resulted over the many and various subsidiary corporations to be a combination or conspiracy in restraint of trade, in violation of the 1st section of the act, [and] also to be an attempt to monopolize and monopolization bringing about a perennial violation of the 2d section. . . .

[W]e think no disinterested mind can survey the period in question without being irresistibly driven to the conclusion that the very genius for commercial development and organization which it would seem was manifested from the beginning soon begot an intent and purpose to exclude others which was frequently manifested by acts and dealings wholly inconsistent with the theory that they were made with the single conception of advancing the development of business power by usual methods, but which, on the contrary, necessarily involved the intent to drive others from the field and to exclude them from their right to trade, and thus accomplish the mastery which was the end in view. And, considering the period from the date of the trust agreements of 1879 and 1882, up to the time of the expansion of the New Jersey corporation, the gradual extension of the power over the commerce in oil which ensued, the decision of the supreme court of Ohio, the tardiness or reluctance in conforming to the commands of that decision, the methods first adopted and that which finally culminated in the plan of the New Jersey corporation, all additionally serve to make manifest the continued existence of the intent which we have previously indicated, and which, among other things, impelled the expansion of the New Jersey corporation. The exercise of the power which resulted from that organization fortifies the foregoing conclusions, since the development which came, the acquisition here and there which ensued of every efficient means by which competition could have been asserted, the slow but resistless methods which followed by which means of transportation were absorbed and brought under control, the system of marketing which was adopted by which the country was divided into districts and the trade in each district in oil was turned over to a designated corporation within the combination, and all others were excluded, all lead the mind up to a conviction of a purpose and intent which we think is so certain as practically to cause the subject not to be within the domain of reasonable contention.

The inference that no attempt to monopolize could have been intended, and that no monopolization resulted from the acts complained of, since it is established that a very small percentage of the crude oil produced was controlled by the combination, is unwarranted. As substantial power over the crude product was the inevitable result of the absolute control which existed over the refined product, the monopolization of the one carried with it the power to control the other. . . .

We are thus brought to the last subject which we are called upon to consider, viz.:

Fourth. The remedy to be administered.

It may be conceded that ordinarily where it was found that acts had been done in violation of the statute, adequate measure of relief would result from restraining the doing of such acts in the future. But in a case like this, where the condition which has been brought about in violation of the statute, in and of itself is not only a continued attempt to monopolize, but also a monopolization, the duty to enforce the statute requires the application of broader and more controlling remedies. As penalties which are not authorized by law may not be inflicted by judicial authority, it follows that to meet the situation with which we are confronted, the application of remedies two-fold in character becomes essential: 1st. To forbid the doing in the future of acts like those which we have found to have been done in the past which would be violative of the statute. 2d. The exertion of such measure of relief as will effectually dissolve the combination found to exist in violation of the statute, and thus neutralize the extension and continually operating force which the possession of the power unlawfully obtained has brought and will continue to bring about. . . .

The court below, by virtue of 1, 2, and 4 of its decree, . . . adjudged that the New Jersey corporation, in so far as it held the stock of the various corporations recited in 2 and 4 of the decree, or controlled the same, was a combination in violation of the 1st section of the act, and an attempt to monopolize or a monopolization contrary to the 2d section of the act. It commanded the dissolution of the combination, and therefore in effect directed the transfer by the New Jersey corporation back to the stockholders of the various subsidiary corporations entitled to the same of the stock which had been turned over to the New Jersey company in exchange for its stock. To make this command effective, 5 of the decree forbade the New Jersey corporation from in any form or manner exercising any ownership or exerting any power directly or indirectly in virtue of its apparent title to the stocks of the subsidiary corporations, and prohibited those subsidiary corporations from paying any dividends to the New Jersey corporations, or doing any act which would recognize further power in that company, except to the extent that it was necessary to enable that company to transfer the stock. So far as the owners of the stock of the subsidiary corporations and the corporations themselves were concerned after the stock had been transferred, 6 of the decree enjoined them from in any way conspiring or combining to violate the act, or to monopolize or attempt to monopolize in virtue of their ownership of the stock transferred to them, and prohibited all agreements between the subsidiary corporations or other stockholders in the future, tending to produce or bring about further violations of the act. . . .

Our conclusion is that the decree below was right and should be affirmed. . . .

And it is so ordered.

APPENDIX C

JOHN DOE I, ET AL., V. UNOCAL CORP., ET AL., AND JOHN ROE III, ET AL., V. UNOCAL CORP., ET AL., BC 237980 AND BC 237679 (2002)

[Note: This court case has been excerpted here. Footnotes and most citations, as well as other material, have been omitted.]

Superior Court of California, County of Los Angeles
Ruling on Unocal Defendants' Motion for Summary Judgment based on: (1) absence of vicarious liability; and (2) failure to join indispensable parties
Hearing date: 6/3/02
Ruling date: 6/10/02

After considering the moving, opposing and reply papers and the arguments of counsel at the hearing, the court now rules as follows:

Unocal Defendants' Motion for Summary Judgment Based on Absence of Vicarious Liability is DENIED.

Unocal Corporation, Union Oil Company of California, John Imle and Roger C. Beach ("Unocal") move for summary judgment, pursuant to Code of Civil Procedure section 437(c), on all Plaintiffs' claims on the grounds that the Unocal defendants have no liability for the alleged tortious acts of individuals in the Myanmar [Burma] military that give rise to the complaints because: (1) the defendants' liability derives solely from the defendant corporations' status as indirect minority shareholders in a corporation that carried out pipeline construction with which the tortious acts are associated; and (2) the undisputed facts and applicable law preclude plaintiffs from recovering under any theory of joint, vicarious or third-party liability alleged in the complaints. Alternatively, Unocal defendants move for summary judgment on all plaintiffs' claims on the grounds that the military and government of Myanmar and TOTAL [the French oil company that was the majority shareholder in the Yadana pipeline project] are indispensable parties who cannot be joined in this action, and without whom relief cannot be granted without prejudice to the Unocal defendants.

Unocal bases this motion on Code of Civil Procedure § 437c, contending that there are no triable issues of material fact and Unocal is entitled to summary judgment as a matter of law.

A defendant is entitled to summary judgment as to each cause of action where one or more of the elements cannot be established. (Code Civ. Proc., § 437c(a).) The party moving for summary judgment bears the initial burden of production to make a prima facie showing of the nonexistence of any triable issue of material fact. The burden shifts to plaintiff when a summary judgment motion prima facie justifies a judgment for the defendant. In moving for summary judgment, a defendant "has met his burden of showing that a cause of action has no merit if he has shown that one or more elements of the cause of action . . . cannot be established, or that there is a complete defense to that cause of action. Once the defendant . . . has met that burden, the burden shifts to the plaintiff . . . to show that a triable issue of one or more material facts exists as to that cause of action or a defense thereto. The plaintiff . . . may not rely upon the mere allegations or denials of his pleadings to show that a triable issue of material fact exists but, instead, must set forth the specific facts showing that a triable issue of material fact exists as to that cause of action or a defense thereto." (*Aguilar v. Atlantic Richfield Co.* (2001), 25 Cal.4th at p. 849, internal quotations omitted.)

The defendant need not conclusively negate such an element, but need only show that the plaintiff does not possess, and cannot reasonably obtain, needed evidence. Once the defendant carries this burden, the burden shifts to plaintiff to make a prima facie showing of the existence of a triable issue of material fact. There is a triable issue of material fact if the evidence would allow a reasonable trier of fact to find the underlying fact in favor of the party opposing the motion in accordance with the applicable standard of proof. . . .

1. CHOICE OF LAW

Unocal argues that Myanmar law applies to plaintiffs' claims for relief. . . . The "snippets and portions" of Myanmar law cited by [Nyein] Kyaw [an expert witness called by Unocal] are not inadequate to identify a conflict for a choice of law analysis and are an insufficient basis on which this Court could decide the complex issues presented by these motions. Therefore, this Court will apply California law.

2. CREATION OF A "YADANA PROJECT JOINT VENTURE"

The terms of the MOU and PSC

Defendants argue that Plaintiffs' theory of vicarious liability fails because the onshore pipeline was built by an independent corporation, not a joint venture with the Myanmar government. In reply, Plaintiffs claim that the Petroleum Production Joint Venture covers "exploitation of natural gas and oil in the Andaman Sea and the construction of a pipeline through the Tenasserim region of Burma." (Doe Complaint, ¶¶ 33–35; Roe Complaint, ¶ 14.)

The existence of a joint venture is an issue of contract. . . .

Defendant's contentions

The creation of the joint venture relationship and the formation of MGTC [Moattama Gas Transportation Company, the company that carried out the Yadana project] are governed by two contracts, the Memorandum of Understanding (MOU) and the Production Sharing Contract (PSC). The express terms of the contracts limit the operations of the Petroleum Production Joint Venture to offshore exploration and development and grant MGTC exclusive authority to conduct pipeline construction and operation. . . .

Plaintiffs' evidence to the contrary

In support of their theory of a larger "Yadana Project Joint Venture," Plaintiffs also rely on provisions of the PSC and MOU. Plaintiffs argue that the scope of the joint venture, as established by these contracts and the intent of the parties as reflected by their conduct, includes onshore activities, and the contracts tie the exploitation of the Mottama gas field to the construction of the pipeline. . . .

Plaintiffs' evidence also shows substantial onshore activities occurred during the two year period between the signing of the MOU and PSC and the formation of MGTC, inconsistent with Unocal's assertion that MGTC was solely responsible for all ground activities. . . .

There is a triable issue of material fact if the evidence would allow a reasonable trier of fact to find the underlying fact in favor of the party opposing the motion in accordance with the applicable standard of proof. Viewed as a whole, Plaintiffs' evidence regarding the contractual language and the intent of the parties, as reflected by their conduct, would allow a reasonable trier of fact to make a finding that would support Plaintiffs' theory of a larger "Yadana Project Joint Venture."

3. WAS MGTC THE ALTER EGO OF THE LARGER "JOINT VENTURE"?

The corporate form may be disregarded where "(1) there is such a unity of interest and ownership between the corporation and the individual or organization controlling it that their separate personalities no longer exist, and (2) failure to disregard the corporate identity would sanction a fraud or promote injustice." (*Webber v. Inland Empire Investments, Inc.* (1999) 74 Cal.App.4th 884, 899.) Relevant factors include the use of a corporation for a single venture, control of day-to-day operations, the commingling of funds and other assets, the disregard of legal formalities, treatment by parents of corporate assets as its own, and the sharing of offices and employees. The essence of the alter ego doctrine is that the parent controls the subsidiary to "such a degree as to render the latter the mere instrumentality of the former." (*Calvert v. Huckins* (E.D.Cal. 1995) 875 F.Supp. 674, 678.) . . .

However, "[a]lter ego is an extreme remedy, sparingly used." (*Sonora Diamond Corp. v. Superior Court* (2000) 83 Cal.App.4th 523, 539.) . . .

Defendants argue that there is no community of interest because the joint venture does not have an ownership interest in MGTC. However, Plaintiffs' unity of interest theory is that the joint venturers are the shareholders of MGTC. . . .

As to the requirement of fraud or injustice, Plaintiffs assert that MGTC was inadequately capitalized at formation, and that Unocal was hiding behind MGTC knowing that tortious conduct was occurring. . . .

Plaintiffs' evidence regarding their alter ego theory would allow a reasonable trier of fact to make a finding that would support Plaintiffs' theory and is sufficient to survive summary judgment.

Other grounds to disregard MGTC's corporate form

Plaintiffs also argue that where a joint venture chooses to conduct its activities through a corporation, courts will summarily disregard the corporate form, contending that the activities of the corporation are those of the joint venture and the joint venture remains liable. . . . [However, p]iercing the corporate veil of a corporation established by a joint venture is not so simple as Plaintiffs would have. Plaintiffs' cases say that formation of the corporation does not preclude a finding of joint venture—however, the cases still apply the alter ego tests to determine whether liability is proper.

4. LIABILITY OF THE "JOINT VENTURE" FOR THE ACTS OF THE MILITARY

a. Joint venture

Plaintiffs argue that Unocal is liable for the acts of the military because the Government of Burma was its joint venturer. A party is liable for the acts of its joint venturer. Assuming that a larger "Yadana Project Joint Venture" existed, there is no basis for imputing the acts of the Burmese military to MOGE [Myanma Oil & Gas Enterprise, a company owned by the Myanmar government that was part of the joint venture] and then through MOGE to MGTC.

Plaintiffs state that it "is doubtful that MOGE ever operates other than as the agent of the military," but cite no evidence relevant to this proposition. The evidence Plaintiffs cite to support imputing liability from the Myanmar military to MOGE include: (1) according to Lipman, the PSC gas rights were given to the joint venturer by the government "through MOGE" (Lipman, p. 111); (2) the negotiations for the project were led on the Burmese side by a military officer, Commander TinTun, the director of the Energy and Planning Department; (3) MOGE detailed how the contractor would calculate taxable income and granted tax exemptions. . .; (4) MOGE undertook to provide security, meant to be carried out by the military; (5) the PSC mandated payment in U.S. dollars. These facts would not allow a reasonable trier of fact to infer that liability passed vicariously from the Myanmar military to MOGE, the governmental entity responsible for administering oil and gas rights.

b. Sovereign immunity

Unocal argues that to the extent that any joint venture with the Myanmar military existed, the sovereign immunity of the Myanmar government would extend to Unocal. The defense of sovereign immunity is personal to the government of Myanmar; Unocal may not benefit from the sovereign immunity which protects MOGE and the government of Myanmar from liability. . . .

c. Agency/Independent contractor theories of vicarious liability

Plaintiffs argue, in the alternative, that if the Burmese government was not a participant in the joint venture, the joint venturers are still liable as principals for the acts of the army. A principal is liable for the intentional torts of his agent committed within the scope of the employment. This includes acts committed by police officers hired to provide security. Whether an agent's act is within the scope of employment depends on whether the act was foreseeable by the principal. This is ordinarily a jury issue.

To authorize an agent, all that is required is conduct by each party manifesting acceptance of a relationship where one party is to perform work for the other under the latter's direction. . . . Failure to discharge an employee who the principal knows has committed a wrongful act is evidence of ratification. Passive acceptance constitutes implied ratification.

Plaintiffs' first argument is that under the PSC and the MGTC/MOGE contracts, that the military was responsible for security for the project. Plaintiffs' evidence shows that in exchange for security, the Project provided food, money and medical assistance for the military assigned to the project.

Plaintiffs' second argument is that if the military's role in the project was not contractual, then the military was an agent of the joint venture. Plaintiff offers some evidence tending to show that the joint venture did, to some extent, direct the operations of the military. Plaintiffs contend that the joint venture ratified the agency relationship by accepting and retaining the benefits of the Myanmar military's security and infrastructure operations. (See *Doe, supra,* 110 F.Supp.2d at p. 1310 ["The evidence does suggest that Unocal knew that forced labor was being utilized and that the Joint Venturers benefitted from the practice"].)

Finally, plaintiffs argue that if the army was not an agent of the joint venture, then it was an independent contractor. Where an employer lacks the legal right to control the employee's activities, the employee is an independent contractor. Plaintiffs present evidence that the joint venture directed the army to accomplish certain tasks, including security and the preparation of roads and helipads. A party is liable for the acts of an independent contractor where the employer should have recognized the harm as likely given the methods adopted by the contractor, or where the employer negligently selected the contractor. The joint venture knew harm was likely, given the military's human rights record.

Because Plaintiffs' evidence would allow a reasonable trier of fact to find that the military was contractually responsible for security, or that the military was an agent or independent contractor hired by the joint venture, sufficient evidence exists to allow plaintiffs to proceed on their independent contractor and agency theories. Summary judgment is not warranted on these grounds.

5. ARE MOGE, TOTAL AND MGTC INDISPENSABLE PARTIES UNDER CODE OF CIVIL PROCEDURE SECTION 389?

Defendants assert that MOGE, TOTAL and MGTC are indispensable parties to this action. . . .

The joinder of these parties is infeasible, as the court does not have personal jurisdiction over these parties, and the Myanmar government enjoys sovereign immunity. However, the issue of whether MOGE, TOTAL and MGTC are indispensable parties has already been litigated between these parties. In *Doe v. Unocal Corp.* (C.D.Cal. 1997) 963 F.Supp. 880, 889 and *NCGUB v. Unocal, Inc.* (C.D.Cal. 1997) 176 F.R.D. 329, 358, the court held that the absence of MOGE and the government of Myanmar would not preclude plaintiffs from obtaining the compensatory relief they requested. Because the parties and the applicable law are identical, Unocal is collaterally estopped from relitigating this issue. . . .

6. IS SUMMARY JUDGMENT WARRANTED?

Unocal brings this motion for summary judgment only under Code of Civil Procedure section 437c. "Any party may move for summary judgment in any action or proceeding if it is contended that the action has no merit or that there is no defense to the action or proceeding." (Code Civ. Proc, § 437c, subd. (a).) "The motion for summary judgment shall be granted if all the papers submitted show that there is no triable issue as to any material fact and that the moving party is entitled to a judgment as a matter of law." (Code Civ. Proc., § 437c, subd. (c).) Summary judgment is denied for two reasons: (1) triable issues of fact exist as to plaintiffs' theories of vicarious liability, as stated above; and (2) even if defendants were entitled to summary adjudication on the vicarious liability issues, plaintiffs' causes of action for violations of the California Constitution, Business & Professions Code section 17200, and unjust enrichment survive. For those reasons, Unocal is not entitled to judgment as a matter of law, and Unocal's motion is denied.

In Sum:
Unocal Defendants' Motion for Summary Judgment Based on Absence of Vicarious Liability is DENIED.

IT IS SO ORDERED.

Dated: 6/10/02
Victoria Gerrard Chaney
Judge

APPENDIX D

Jose Francisco Sosa v. Humberto Alvarez-Machain et al., 03-339 (2004)

[Note: This court decision has been excerpted here. Most footnotes and citations, as well as other material, have been omitted.]

On writs of certiorari to the United States Court of Appeals for the Ninth Circuit
[June 29, 2004]

Justice Souter delivered the opinion of the Court.

The two issues are whether respondent Alvarez-Machain's allegation that the Drug Enforcement Administration instigated his abduction from Mexico for criminal trial in the United States supports a claim against the Government under the Federal Tort Claims Act (FTCA or Act), 28 U. S. C. §1346(b)(1), §§2671-2680, and whether he may recover under the Alien Tort Statute (ATS), 28 U. S. C. §1350. We hold that he is not entitled to a remedy under either statute.

I

We have considered the underlying facts before. In 1985, an agent of the Drug Enforcement Administration (DEA), Enrique Camarena-Salazar, was captured on assignment in Mexico and taken to a house in Guadalajara, where he was tortured over the course of a 2-day interrogation, then murdered. Based in part on eyewitness testimony, DEA officials in the United States came to believe that respondent Humberto Alvarez-Machain (Alvarez), a Mexican physician, was present at the house and acted to prolong the agent's life in order to extend the interrogation and torture.

In 1990, a federal grand jury indicted Alvarez for the torture and murder of Camarena-Salazar, and the United States District Court for the Central District of California issued a warrant for his arrest. The DEA asked the Mexican Government for help in getting Alvarez into the United States, but when the requests and negotiations proved fruitless, the DEA approved a plan to hire Mexican nationals to seize Alvarez and bring him to the United States for trial. As so planned, a group of Mexicans, including petitioner Jose Francisco Sosa, abducted Alvarez from his house, held him overnight in a motel, and brought him by private plane to El Paso, Texas, where he was arrested by federal officers.

Once in American custody, Alvarez moved to dismiss the indictment on the ground that his seizure was "outrageous governmental conduct," *Alvarez-Machain*, 504 U. S., at 658, and violated the extradition treaty between the United States and Mexico. The District Court agreed, the Ninth Circuit affirmed, and we reversed, holding that the fact of Alvarez's forcible seizure did not affect the jurisdiction of a federal court. The case was tried in 1992, and ended at the close of the Government's case, when the District Court granted Alvarez's motion for a judgment of acquittal.

In 1993, after returning to Mexico, Alvarez began the civil action before us here. He sued Sosa, Mexican citizen and DEA operative Antonio Garate-Bustamante, five unnamed Mexican civilians, the United States, and four DEA agents. So far as it matters here, Alvarez sought damages from the United States under the FTCA, alleging false arrest, and from Sosa under the ATS, for a violation of the law of nations. The former statute authorizes suit "for . . . personal injury . . . caused by the negligent or wrongful act or omission of any employee of the Government while acting within the scope of his office or employment." 28 U. S. C. §1346(b)(1). The latter provides in its entirety that "[t]he district courts shall have original jurisdiction of any civil action by an alien for a tort only, committed in violation of the law of nations or a treaty of the United States." §1350.

The District Court granted the Government's motion to dismiss the FTCA claim, but awarded summary judgment and $25,000 in damages to Alvarez on the ATS claim. A three-judge panel of the Ninth Circuit then affirmed the ATS judgment, but reversed the dismissal of the FTCA claim.

A divided en banc court came to the same conclusion. As for the ATS claim, the court called on its own precedent, "that [the ATS] not only provides federal courts with subject matter jurisdiction, but also creates a cause of action for an alleged violation of the law of nations." *Id.*, at 612. The Circuit then relied upon what it called the "clear and universally recognized norm prohibiting arbitrary arrest and detention," *id.*, at 620, to support the conclusion that Alvarez's arrest amounted to a tort in violation of international law. On the FTCA claim, the Ninth Circuit held that, because "the DEA had no authority to effect Alvarez's arrest and detention in Mexico," *id.*, at 608, the United States was liable to him under California law for the tort of false arrest.

We granted certiorari in these companion cases to clarify the scope of both the FTCA and the ATS. We now reverse in each. . . .

[discussion of FTCA claim omitted]

III

Alvarez has also brought an action under the ATS against petitioner, Sosa, who argues (as does the United States supporting him) that there is no relief under the ATS because the statute does no more than vest federal courts with jurisdiction, neither creating nor authorizing the courts to recognize any particular right of action without further congressional action. Although we agree the statute is in terms only jurisdictional, we think that at the time of enactment the jurisdiction enabled federal courts to hear claims in a very limited category defined by the law of nations and recognized at common law. We do not believe, however, that the limited, implicit sanction to entertain the handful of international law *cum* common law claims understood in 1789 should be taken as authority to recognize the right of action asserted by Alvarez here.

A

Judge Friendly called the ATS a "legal Lohengrin," *IIT* v. *Vencap, Ltd.*, 519 F. 2d 1001, 1015 (CA2 1975); "no one seems to know whence it came," *ibid.*, and for over 170 years after its enactment it provided jurisdiction in only one case. The first Congress passed it as part of the Judiciary Act of 1789, in providing that the new federal district courts "shall also have cognizance, concurrent with the courts of the several States, or the circuit courts, as the case may be, of all causes where an alien sues for a tort only in violation of the law of nations or a treaty of the United States." Act of Sept. 24, 1789, ch. 20, §9*(b)*, 1 Stat. 79.

The parties and *amici* here advance radically different historical interpretations of this terse provision. Alvarez says that the ATS was intended not simply as a jurisdictional grant, but as authority for the creation of a new cause of action for torts in violation of international law. We think that reading is implausible. As enacted in 1789, the ATS gave the district courts "cognizance" of certain causes of action, and the term bespoke a grant of jurisdiction, not power to mold substantive law. . . . The fact that the ATS was placed in §9 of the Judiciary Act, a statute otherwise exclusively concerned with federal-court jurisdiction, is itself support for its strictly jurisdictional nature. Nor would the distinction between jurisdiction and cause of action have been elided by the drafters of the Act or those who voted on it. . . . In sum, we think the statute was intended as jurisdictional in the sense of addressing the power of the courts to entertain cases concerned with a certain subject.

But holding the ATS jurisdictional raises a new question, this one about the interaction between the ATS at the time of its enactment and the ambient law of the era. Sosa would have it that the ATS was stillborn because there could be no claim for relief without a further statute expressly authorizing adoption of causes of action. *Amici* professors of federal jurisdiction and legal history take a different tack, that federal courts could entertain claims once the jurisdictional grant was on the books, because torts in violation of the law of nations would have been recognized within the common law of the time. We think history and practice give the edge to this latter position.

1

"When the *United States* declared their independence, they were bound to receive the law of nations, in its modern state of purity and refinement." *Ware* v. *Hylton*, 3 Dall. 199, 281 (1796) (Wilson, J.). In the years of the early Republic, this law of nations comprised two principal elements, the first covering the general norms governing the behavior of national states with each other. . . . This aspect of the law of nations . . . occupied the executive and legislative domains, not the judicial. . . .

The law of nations included a second, more pedestrian element, however, that did fall within the judicial sphere, as a body of judge-made law regulating the conduct of individuals situated outside domestic boundaries and consequently carrying an international savor. To Blackstone, the law of nations in this sense was implicated "in mercantile questions, . . . [and] in all disputes relating to prizes, to shipwrecks, to hostages, and ransom bills." *Id.*, at 67. . . .

There was, finally, a sphere in which these rules binding individuals for the benefit of other individuals overlapped with the norms of state relationships. Blackstone referred to it when he mentioned three specific offenses against the law of nations addressed by the criminal law of England: violation of safe conducts, infringement of the rights of ambassadors, and piracy. 4 Commentaries 68. An assault against an ambassador, for example, impinged upon the sovereignty of the foreign nation and if not adequately redressed could rise to an issue of war. It was this narrow set of violations of the law of nations, admitting of a judicial remedy and at the same time threatening serious consequences in international affairs, that was probably on the minds of the men who drafted the ATS with its reference to tort. . . .

3

Although Congress modified the draft of what became the Judiciary Act, it made hardly any changes to the provisions on aliens, including what became the ATS. There is no record of congressional discussion about private actions that might be subject to the jurisdictional provision, or about any need for further legislation to create private remedies; there is no record even of debate on the section. . . . [D]espite considerable scholarly attention, it is fair to say that a consensus understanding of what Congress intended has proven elusive.

Still, the history does tend to support two propositions. First, there is every reason to suppose that the First Congress did not pass the ATS as a jurisdictional convenience to be placed on the shelf for use by a future Congress or state legislature that might, some day, authorize the creation of causes of action or itself decide to make some element of the law of nations actionable for the benefit of foreigners. . . .

The second inference to be drawn from the history is that Congress intended the ATS to furnish jurisdiction for a relatively modest set of actions alleging violations of the law of nations. Uppermost in the legislative mind appears to have been offenses against ambassadors, violations of safe conduct were probably understood to be actionable, and individual actions arising out of prize captures and piracy may well have also been contemplated. But the common law appears to have understood only those three of the hybrid variety as definite and actionable, or at any rate, to have assumed only a very limited set of claims. As Blackstone had put it, "offences against this law [of nations] are principally incident to whole states or nations," and not individuals seeking relief in court. 4 Commentaries 68. . . .

In sum, although the ATS is a jurisdictional statute creating no new causes of action, the reasonable inference from the historical materials is that the statute was intended to have practical effect the moment it became law. The jurisdictional grant is best read as having been enacted on the understanding that the common law would provide a cause of action for the modest number of international law violations with a potential for personal liability at the time.

IV

We think it is correct, then, to assume that the First Congress understood that the district courts would recognize private causes of action for certain torts in violation of the law of nations, though we have found no basis to suspect Congress had any examples in mind beyond those torts corresponding to Blackstone's three primary offenses: violation of safe conducts, infringement of the rights of ambassadors, and piracy. We assume, too, that no development in the two centuries from the enactment of §1350 to the birth of the modern line of cases beginning with *Filartiga* v. *Peña-Irala*, 630 F. 2d 876 (CA2 1980), has categorically precluded federal courts from recognizing a claim under the law of nations as an element of common law; Congress has not in any relevant way amended §1350 or limited civil common law power by another statute. Still, there are good reasons for a restrained conception of the discretion a federal court should exercise in considering a new cause of action of this kind. Accordingly, we think courts should require any claim based on the present-day law of nations to rest on a norm of international character accepted by the civilized world and defined with a specificity comparable to the features of the 18th-century paradigms we have recognized. This requirement is fatal to Alvarez's claim.

A

A series of reasons argue for judicial caution when considering the kinds of individual claims that might implement the jurisdiction conferred by the early statute. First, the prevailing conception of the common law has changed since 1789 in a way that counsels restraint in judicially applying internationally generated norms. When §1350 was enacted, the accepted conception was of the common law as "a transcendental body of law outside of any particular State but obligatory within it unless and until changed by statute." *Black and White Taxicab & Transfer Co.* v. *Brown and Yellow Taxicab & Transfer Co.*, 276 U. S. 518, 533 (1928) (Holmes, J., dissenting). Now, however, in most cases where a court is asked to state or formulate a common law principle in a new context, there is a general understanding that the law is not so much found or discovered as it is either made or created. . . . [A] judge deciding in reliance on an international norm will find a substantial element of discretionary judgment in the decision.

Second, along with, and in part driven by, that conceptual development in understanding common law has come an equally significant rethinking of the role of the federal courts in making it. . . . [T]he general practice has been to look for legislative guidance before exercising innovative authority over substantive law. It would be remarkable to take a more aggressive role in exercising a jurisdiction that remained largely in shadow for much of the prior two centuries.

Third, this Court has recently and repeatedly said that a decision to create a private right of action is one better left to legislative judgment in the great majority of cases. . . . [Especially,] the possible collateral consequences of making international rules privately actionable argue for judicial caution.

Fourth, the subject of those collateral consequences is itself a reason for a high bar to new private causes of action for violating international law, for the potential implications for the foreign relations of the United States of recognizing such causes should make courts particularly wary of impinging on the discretion of the Legislative and Executive Branches in managing foreign affairs. It is one thing for American courts to enforce constitutional limits on our own State and Federal Governments' power, but quite another to consider suits under rules that would go so far as to claim a limit on the power of foreign governments over their own citizens, and to hold that a foreign government or its agent has transgressed those limits. Yet modern international law is very much concerned with just such questions, and apt to stimulate calls for vindicating private interests in §1350 cases. Since many attempts by federal courts to craft remedies for the violation of new norms of international law would raise risks of adverse foreign policy consequences, they should be undertaken, if at all, with great caution. Cf. *Tel-Oren* v. *Libyan Arab Republic*, 726 F. 2d 774, 813 (CADC 1984) (Bork, J., concurring) (expressing doubt that §1350 should be read to require "our courts [to] sit in judgment of the conduct of foreign officials in their own countries with respect to their own citizens").

The fifth reason is particularly important in light of the first four. We have no congressional mandate to seek out and define new and debatable violations of the law of nations, and modern indications of congressional understanding of the judicial role in the field have not affirmatively encouraged greater judicial creativity. It is true that a clear mandate appears in the Torture Victim Protection Act of 1991, 106 Stat. 73, providing authority that "establish[es] an unambiguous and modern basis for" federal claims of torture and extrajudicial killing, H. R. Rep. No. 102.367, pt. 1, p. 3 (1991). But that affirmative authority is confined to specific subject matter, and although the legislative history includes the remark that §1350 should "remain intact to permit suits based on other norms that already exist or may ripen in the future into rules of customary international law," *id.*, at 4, Congress as a body has done nothing to promote such suits. Several times, indeed, the Senate has expressly declined to give the federal courts the task of interpreting and applying international human rights law, as when its ratification of the International Covenant on Civil and Political Rights declared that the substantive provisions of the document were not self-executing. 138 Cong. Rec. 8071 (1992).

B

These reasons argue for great caution in adapting the law of nations to private rights. JUSTICE SCALIA, *post*, p. 1 (opinion concurring in part and concurring in judgment) concludes that caution is too hospitable, and a word is in order to summarize where we have come so far and to focus our difference with him on whether some norms of today's law of nations may ever be recognized legitimately by federal courts in the absence of congressional action beyond §1350. All Members of the Court agree that §1350 is only jurisdictional. We also agree, or at least JUSTICE SCALIA does not dispute, that the jurisdiction was originally understood to be available to enforce a small number of international norms that a federal court could properly recognize as within the common law enforceable without further statutory authority. JUSTICE SCALIA concludes, however, that two subsequent developments should be understood to preclude federal courts from recognizing any further international norms as judicially enforceable today, absent further congressional action. As described before, we now tend to understand common law not as a discoverable reflection of universal reason but, in a positivistic way, as a product of human choice. And we now adhere to a conception of limited judicial power first expressed in reorienting federal diversity jurisdiction, see *Erie R. Co.* v. *Tompkins*, 304 U. S. 64 (1938), that federal courts have no authority to derive "general" common law.

Whereas JUSTICE SCALIA sees these developments as sufficient to close the door to further independent judicial recognition of actionable international norms, other considerations persuade us that the judicial power should be exercised on the understanding that the door is still ajar subject to vigilant doorkeeping, and thus open to a narrow class of international norms today. *Erie* did not in terms bar any judicial recognition of new substantive rules, no matter what the circumstances, and post-*Erie* understanding has identified limited enclaves in which federal courts may derive some substantive law in a common law way. For two centuries we have affirmed that the domestic law of the United States recognizes the law of nations. . . . It would take some explaining to say now that federal courts must avert their gaze entirely from any international norm intended to protect individuals. . . .

The position we take today has been assumed by some federal courts for 24 years, ever since the Second Circuit decided *Filartiga* v. *Peña-Irala*, and for practical purposes the point of today's disagreement has been focused since the exchange between Judge Edwards and Judge Bork in *Tel-Oren* v. *Libyan Arab Republic*, 726 F. 2d 774 (CADC 1984), Congress, however, has not only expressed no disagreement with our view of the proper exercise of the judicial power, but has responded to its most notable instance by enacting legislation supplementing the judicial determination in some detail [in the form of the Torture Victim Protection Act].

While we agree with JUSTICE SCALIA to the point that we would welcome any congressional guidance in exercising jurisdiction with such obvious potential to affect foreign relations, nothing Congress has done is a reason for us to shut the door to the law of nations entirely. . . .

C

We must still, however, derive a standard or set of standards for assessing the particular claim Alvarez raises, and for this case it suffices to look to the historical antecedents. Whatever the ultimate criteria for accepting a cause of action subject to jurisdiction under §1350, we are persuaded that federal courts should not recognize private claims under federal common law for violations of any international law norm with less definite content and acceptance among civilized nations than the historical paradigms familiar when §1350 was enacted. . . . This limit upon judicial recognition is generally consistent with the reasoning of many of the courts and judges who faced the issue before it reached this Court. See *Filartiga, supra*, at 890 ("[F]or purposes of civil liability, the torturer has become, like the pirate and slave trader before him, *hostis humani generis*, an enemy of all mankind"); *Tel-Oren, supra*, at 781 (Edwards, J., concurring) (suggesting that the "limits of section 1350's reach" be defined by "a handful of heinous actions, each of which violates definable, universal and obligatory norms"); see also *In re Estate of Marcos Human Rights Litigation*, 25 F. 3d 1467, 1475 (CA9 1994) ("Actionable violations of international law must be of a norm that is specific, universal, and obligatory"). And the determination whether a norm is sufficiently definite to support a cause of action [fn. 20] should (and, indeed, inevitably must) involve an element of judgment about the practical consequences of making that cause available to litigants in the federal courts.

Thus, Alvarez's detention claim must be gauged against the current state of international law. . . .

To begin with, Alvarez cites two well-known international agreements that, despite their moral authority, have little utility under the standard set out in this opinion. He says that his abduction by Sosa was an "arbitrary arrest" within the meaning of the Universal Declaration of Human Rights (Declaration), G. A. Res. 217A (III), U. N. Doc. A/810 (1948). And he

[20] A related consideration is whether international law extends the scope of liability for a violation of a given norm to the perpetrator being sued, if the defendant is a private actor such as a corporation or individual. Compare *Tel-Oren* v. *Libyan Arab Republic*, 726 F. 2d 774, 791.795 (CADC 1984) (Edwards, J., concurring) (insufficient consensus in 1984 that torture by private actors violates international law), with *Kadic* v. *Karadzic*, 70 F. 3d 232, 239.241 (CA2 1995) (sufficient consensus in 1995 that genocide by private actors violates international law).

traces the rule against arbitrary arrest not only to the Declaration, but also to article nine of the International Covenant on Civil and Political Rights (Covenant), Dec. 19, 1996, 999 U. N. T. S. 171, 22 to which the United States is a party, and to various other conventions to which it is not. But the Declaration does not of its own force impose obligations as a matter of international law. . . . [A]lthough the Covenant does bind the United States as a matter of international law, the United States ratified the Covenant on the express understanding that it was not self-executing and so did not itself create obligations enforceable in the federal courts. Accordingly, Alvarez cannot say that the Declaration and Covenant themselves establish the relevant and applicable rule of international law. He instead attempts to show that prohibition of arbitrary arrest has attained the status of binding customary international law.

Here, it is useful to examine Alvarez's complaint in greater detail. As he presently argues it, the claim does not rest on the cross-border feature of his abduction. Although the District Court granted relief in part on finding a violation of international law in taking Alvarez across the border from Mexico to the United States, the Court of Appeals rejected that ground of liability for failure to identify a norm of requisite force prohibiting a forcible abduction across a border. Instead, it relied on the conclusion that the law of the United States did not authorize Alvarez's arrest, because the DEA lacked extra-territorial authority under 21 U. S. C. §878, and because Federal Rule of Criminal Procedure 4(d)(2) limited the warrant for Alvarez's arrest to "the jurisdiction of the United States." It is this position that Alvarez takes now: that his arrest was arbitrary and as such forbidden by international law not because it infringed the prerogatives of Mexico, but because no applicable law authorized it.

Alvarez thus invokes a general prohibition of "arbitrary" detention defined as officially sanctioned action exceeding positive authorization to detain under the domestic law of some government, regardless of the circumstances. Whether or not this is an accurate reading of the Covenant, Alvarez cites little authority that a rule so broad has the status of a binding customary norm today. He certainly cites nothing to justify the federal courts in taking his broad rule as the predicate for a federal lawsuit, for its implications would be breathtaking. His rule would support a cause of action in federal court for any arrest, anywhere in the world, unauthorized by the law of the jurisdiction in which it took place, and would create a cause of action for any seizure of an alien in violation of the Fourth Amendment. . . . It would create an action in federal court for arrests by state officers who simply exceed their authority; and for the violation of any limit that the law of any country might place on the authority of its own officers to arrest. And all of this assumes that Alvarez could establish that Sosa was acting on behalf of a government when he made the arrest, for otherwise he would need a rule broader still.

Alvarez's failure to marshal support for his proposed rule is underscored by the Restatement (Third) of Foreign Relations Law of the United States (1987), which says in its discussion of customary international human rights law that a "state violates international law if, as a matter of state policy, it practices, encourages, or condones . . . prolonged arbitrary detention." *Id.*, §702. Although the Restatement does not explain its requirements of a "state policy" and of "prolonged" detention, the implication is clear. Any credible invocation of a principle against arbitrary detention that the civilized world accepts as binding customary international law requires a factual basis beyond relatively brief detention in excess of positive authority. Even the Restatement's limits are only the beginning of the enquiry, because although it is easy to say that some policies of prolonged arbitrary detentions are so bad that those who enforce them become enemies of the human race, it may be harder to say which policies cross that line with the certainty afforded by Blackstone's three common law offenses. In any event, the label would never fit the reckless policeman who botches his warrant, even though that same officer might pay damages under municipal law.

Whatever may be said for the broad principle Alvarez advances, in the present, imperfect world, it expresses an aspiration that exceeds any binding customary rule having the specificity we require. Creating a private cause of action to further that aspiration would go beyond any residual common law discretion we think it appropriate to exercise. It is enough to hold that a single illegal detention of less than a day, followed by the transfer of custody to lawful authorities and a prompt arraignment, violates no norm of customary international law so well defined as to support the creation of a federal remedy.

* * *

The judgment of the Court of Appeals is
Reversed.

INDEX

Locators in **boldface** indicate main topics. Locators followed by *c* indicate chronology entries. Locators followed by *b* indicate biographical entries. Locators followed by *g* indicate glossary entries.

Index

Index

Index

Index

Index

Index

Index

Index

Index

Index